Flipons

Alan Herbert has played a leading part in discovering the biological roles for a high-energy form of DNA twisted to the left rather than to the right. Both Z-DNA and the Z-RNA-sensing proteins are critical for protecting hosts against both viruses and cancers. The proteins also play critical roles in the programmed cell death of aging cells. Other types of flipon exist and alter the readout of transcripts from the genome, encoding genetic information by their shape rather than by their sequence. Many of these flipons are within repeat elements that were previously considered to be just genomic junk. Instead, these genetic elements increase the adaptability of cells by flipping DNA conformation. By acting as digital switches, the different flipon types can alter cellular responses without any change to their sequence or any damage to DNA. These highly dynamic structures enable the rapid evolution of multicellular organisms. The junk DNA in repeats also encodes peptide patches that enable the assembly of cellular machines. The intransitive logic involved enhances the chance of an individual surviving a constantly changing environment.

Key Features

- Causes us to rethink how information is encoded in the genome.
- Changes our understanding of how our genome evolved and how we protect ourselves against viruses and cancers while sparing normal cells.
- Shows that high-energy forms of DNA, such as left-handed DNA, exist inside the cell.
- Describes how aging and disease depends on pathways not active in normal tissues.
- Accessible to those in academia and the general public, and speaks to the next generation, encouraging them to find their own path in scientific discovery.

Flipons

The Discovery of Z-DNA and Soft-Wired Genomes

Alan Herbert

CRC Press
Taylor & Francis Group
Boca Raton London New York

CRC Press is an imprint of the
Taylor & Francis Group, an **informa** business

Designed cover image: Alan Herbert

First edition published 2024
by CRC Press
2385 NW Executive Center Drive, Suite 320, Boca Raton FL 33431

and by CRC Press
4 Park Square, Milton Park, Abingdon, Oxon, OX14 4RN

CRC Press is an imprint of Taylor & Francis Group, LLC

ISBN: 9781032732961 (hbk)
ISBN: 9781032731087 (pbk)
ISBN: 9781003463535 (ebk)

DOI: 10.1201/9781003463535

Typeset in Times
by Deanta Global Publishing Services, Chennai, India

Dedication

---

*Men are disturbed not by things, but by the
view which they take of them.*

—**Epictetus**, *The Enchiridion*

Contents

PART II

Preface

The original title of this book was *Flipon Science – The Strange Twists in the Discovery of a Biological Role for Left-handed DNA*. The naming was full of puns. The word "flipons" relates to a significant advance in the way we understand our genome. A flip not only of our DNA but also in what we conceive as possible. The book advances our understanding of how the genome encodes information. The focus is on how the shape of DNA, rather than the sequences, affects the readout of the information in our genes. The twist of DNA to the left rather than to the right affects both health and disease. The realization changes our views of how cells evolve.

The flipons are also a meme for my different roles in these discoveries. Each phase required a change of footwear for the next stage in the journey. The most recent advances were made rather casually. Flipons were adequate foot protection for the final climb over the mountains of data I used from the Human Genome Project. I did not wear a white lab coat, protective footwear, nor strap on safety goggles. The required bits of information were amassed somewhere out there in the ether, waiting to find their final form in the story they revealed.

In the early part of the journey, I did require all that protective equipment and more! The formal description of the story would go like this. "The discovery of a protein that bound to a left-handed DNA helix was achieved following the well-established tenets of rigorous bench science based on stringently controlled experimentation using a multitude of approaches". Then, a historian of science might write "The advances were led by the computational analysis of large datasets accumulated through massive genomic sequencing studies performed on an industrial scale, yielding genetic predictions of such high validity that the hypotheses were rapidly confirmed experimentally".

One editor told me that no one would read a book written as is customary in the scientific literature. That would be a pity as the story is one everyone can relate to regardless of what they do. Who has never been told that what they are doing will never work? Or been given the friendly advice that people might think you are either a fool or on a fool's errand if you continue on your current path? Or, out of "fairness", been given a "choice" that really has just one option? Of course, we no longer live in the age where ultimatums enforce compliance. When given your "choice", how have you responded? Did you ever wonder what would have happened if you had chosen differently? What if you had taken the road Robert Frost called "less traveled"? Or, if, like Neo from the film *The Matrix*, you had taken the red pill instead of the blue one?

This book is based on my recollections. I was lucky enough to work on a problem when it was not known whether there was a solution and to find that one did indeed exist. Of course, as in charting any unexplored waters, there were snatches of smooth sailing swamped by stormy seas. It was not always the collegial collaborations that we collectively conceive as the essence of science. It was not Disneyland and it was not *Mr Roger's Neighborhood*. In life, you can't really go to a pristine wilderness like

Yellowstone and pet the local bears because they look like do "Smokey the Bear" from your childhood. The real bears will bite your arm off. In science, it is no different and the bears do look very professorial, just like Smokey. Nor is it safe to use public opinion polls to judge whether your science is good or bad. Take the case of Antoine Lavoisier who perished during the French Revolution of 1789. He was popular as the chemist credited with the discovery of oxygen, among other contributions to science. As a tax-farmer who bought the right to collect taxes from his neighbors, he tanked in the popularity polls - despite his great discoveries. He met his end at the hands of the mob, along with other less-talented individuals. Apparently, the mob did not like the way they were being farmed: the wisdom of the crowd on full display! Fortunately, my journey ended differently, as you will see.

Another title for this book was *The Flipon Wars*. If there was a war, it was a small-scale guerrilla style conflict. The skirmishes were asymmetrical, with the vast majority of scientists opposing the advances we made. They were in command of the weapons and strategic positions necessary to fund and locate their troops to defend the turf they owned. Their intent was to deny us a foothold. How do you prevail against such odds? Of course, the description is overly dramatic – no one as far as I know was imprisoned or killed for studying flipons. So, it really wasn't a war, but much more interesting as the events played out over many years, allowing for the many plot twists that I will relate to you. It is a story of how science advances. The only thing destroyed were the biases and bad ideas. There were careers unfortunately cut short, but that risk was soon apparent quite early on to everyone playing this science-based version of the *Hunger Games*. Of course, no one believes it would happen to them. But it did.

In the battle to advance scientific discovery, naturally there is a bureaucracy to contend with. Those who manage how research is funded, and grant academic appointments, will, without a doubt, say "Let's fund those things that we expect to be true and publish those findings that fit". While alternative explanations may be fun and show ingenuity, the probability is that almost all will fail or be flawed. That's the stuff students learn not to do if they are to advance their careers.

Being bureaucratic is a burden even for those who run the show. A funding institution like the NIH only wants tables full of statistics that list the new resources created, the publications funded, or other outcomes that they can credit to their programs. The quantitative measures are assembled to show the productivity needed to justify new funding by the politicians, who, predictably, want the voters to know that they care for them so much. Of course, the NIH outcome measures improve when people stop smoking, eat better, reduce salt intake, drink less alcohol, wear sunscreen and just say no. The NIH statistics are further boosted by public health initiatives that clean water, sanitize sewage, improve air quality, eliminate toxins, and remove mutagens from the environment. Of course, the NIH messaging of how vital their role is in the nation's health would be lost if the NIH did not have the anti-vaxxers, the anti-pharma, and the anti-woke to state the issues in a way that the public can clearly understand and vote for or against, which brings us back to the politics of how the NIH funds science.

The financiers are guided by their perceptions of what is possible. That does cause a problem or two. As Yogi Berra noted: "It is difficult to make predictions, especially about the future". Normally, that leaves only one easy solution. Those with money in the game want a sure return on investment within two to five years. Of course, even politicians, who want to appear progressive, press for more risk-taking. Every now and again, the politicians launch a new "War on Cancer" or "A Cancer Moonshot". Who can find fault in that? Unfortunately, those political initiatives have not proven particularly effective either. The same key opinion leaders end up guiding both the politicians and the NIH. Eventually, those selling the mission get back to the business of staying relevant and funded for the next two to five years.

At some point, you will have to answer your critics. Saying nothing is not always an option. Often, the best response to authority is to quote a higher authority. Here are some good examples that will make them ponder what to say next, giving you time to discreetly make an exit. As Max Planck famously noted in 1948, "A new scientific truth does not generally triumph by persuading its opponents and getting them to admit their errors, but rather by its opponents gradually dying out and giving way to a new generation that is raised on it". A similar thought had been expressed in an earlier generation by Francis Darwin, the son of Charles, in 1881. "It is always the case with the best work, that it is misrepresented, and disparaged at first, for it takes a curiously long time for new ideas to become current, and the older men who ought to be capable of taking them in freely, will not do so through prejudice". (quoted in [1]). A similar notion was expressed in a later generation. In 1979, Francis Crick noted that "Lacking evidence we ... become overconfident in the generality of some of our basic ideas" [2]. Remember to leave as people ponder these words in their minds: this conversation rarely ends well. After all, they are your critics. Bulls are color-blind so it really doesn't matter what color the flag is that you are waving at them.

How, then, can you make discoveries when the area of research is not a funding priority and when the majority of your peers who review the grant applications believe the field you work is not relevant? Well, there's the rub. I will give you here my version of Hamlet's "To be or not to be" soliloquy, or at least the one given in my paper that was entitled "To Z or not to Z" [3].

Of course, the experiments need to be funded and someone has to perform them, and carefully. The new, ground-breaking insights will then seamlessly transform into common wisdom that any grade-level child knows, as if it were always so. The process moves more at the pace of a Gandhi spiritual awakening across India rather than with the fanatical pace of a Che Guevara armed insurrection. There is nothing physical to destroy, just laws that have no basis in the natural order of science. Then, finally, the change you wish for happens. It's as if there were a miraculous intervention. It's as if someone said "Let there be light". A hypothesis that was initially considered irrational suddenly becomes irrefutable. You then hear "I always knew that" rather than the previous dismissals of "Why would you ever work on "X"?" (in my case, "X" was Z-DNA). So, where does the light come from, you might properly ask? Read on.

What happens if, despite all the drama and all the pain(s), you are lucky enough to reach your nirvana? In Science, "Victory at last" is not the sound of marching bands. It takes a while for the news of a miracle to spread. Science is, at best, more of a Tom Jones story (from the book, not the singer), where Tom follows where his heart takes him, then lives happily ever after. So, this book is more a chronicle than an epic. It is not *War and Peace* with heroic battles and vanquished armies.

Everyone is taught that new ideas are resisted, and they quote "First, they ignore you, then they laugh at you, then they fight you, then you win" (falsely attributed to Mahatma Gandhi, https://apnews.com/article/archive-fact-checking-231580316). Knowing this does not prepare you for being part of that narrative. The "vainglorious chatterers" Paracelsus described will go silent (quote from *Ten Drugs: How Plants, Powders, and Pills Have Shaped the History of Medicine* by Thomas Hager [2019]). That is, if you are lucky enough to survive long enough to witness the miracle! The only way to let everyone know you are still around and have done something special is to publish and hope that others eventually take notice. Your words will survive way beyond those of your critics. But there again is a challenge. Where can you publish if editors don't think anyone is interested in your work? What happens if an editor does like your work but can't find anyone willing to spend their time reviewing an article of such dubious significance? In both cases, you will receive the standard letter: "Thank you for considering our journal. We receive many more submissions than we can publish…". You need to publish, so persist.

I would like to say that the story I tell is unparalleled in the history of Science. It is not. It is an old adage adorned in more modern garb. We are pretty much cut from the same cloth as our predecessors. Yet, we deny that. We distrust new things and we instinctively dislike those outcomes that are different from what we expect. Like those who went before us, it takes time for us all to accept that change happens. Like them, we prefer our version of the future. Indeed, most of us consider that our IQ is far higher than the average score of 100, so why should others know better than us? Furthermore, we consider ourselves technologically advanced and sophisticated in a way far beyond the capabilities of our ancestors or those who lack our education. Yet, in the same way that you reflexively pick up the phone to call a parent who has just passed on, it takes time to adjust to the new realities. So, the reluctance of others to accept change runs through my story.

Interestingly, some of the worst opponents you will face are those colleagues you work with. Often, others outside the lab turn to them for the "inside scoop". Your compatriots know of all your past failures and are aware of all the flameouts likely to occur, often projected far into the future. They want no association with any of that, "Who me? … No, that has nothing to do with me" or they might say "Remember the time when he …". On the other hand, most people like being on the bandwagon if they can find one on the road to their success. It is just a question of what they will do to get their seat and whether they feel that they are more deserving of the ride than you. But then, you may prefer jazz over gospel, so why not start your own parade? Why not lace up your own marching boots and move to your own rhythm?

I am originally from New Zealand, where I could not do the crazy things I wanted to do, so why not vote with my feet? I mean to say, Edmund Hillary had to leave New Zealand to climb Everest. I came to the United States to find my own mountain to scale. My aim was to reach for the blue sky high above. The challenge that attracted me was to unravel the biology of an unconventional form of DNA, a helix that is twisted to the left, rather than to the right. If that sounds esoteric, then you are correct. But it is no more an esoteric pursuit than climbing Mount Everest, for no other reason than "… because it was there". The issue in both cases is to test how well you measure up. We score by our success but, more often, are judged by our missteps. Unlike mountaineering, failure in science is usually not fatal. Unlike mountaineering, a successful outcome is often not a Kodak moment that can be flashed across the world. There is no peak you can plant a flag on, few pictures, and no captions as simple as "Hillary Tops Mount Everest". Labeling the model of Watson-Crick DNA as "The Secret of Life", although attention grabbing, is not exactly as self-evident as a porn star with a president. However, there is a difference between what sells newspapers and those events that change our collective futures.

Many of our advances as a species have come from our studies of how Nature works. Nature builds intelligent life in ways that continue to surprise, especially knowing that this complexity was built from the simple stuff of the primordial broth. We still have no way to match what we see around us. We have little understanding of how to reverse engineer the energy-efficient solutions Nature has devised. Nature has evolved self-powering, self-replicating, self-repairing, and self-aware systems. They function in a range of extreme environments. The story I relate here adds to our understanding of these outcomes. I describe how Nature found a use for higher-energy forms of DNA, focusing first on the discovery of Z-DNA. I then move on to how the biological function of Z-DNA was revealed. I describe how dynamic DNA and RNA structures, like Z-DNA and Z-RNA, allow a cell to program different outcomes by changing their conformation without changing a single base of DNA sequence. Then I introduce other types of flipons. I examine how a cell builds genetic programs based on flipons. This leads to a discussion of soft-wired genomes, where DNA does not set your destiny. Instead, the information in your genome is read out in many different ways, allowing Nature to solve for survival almost instantly. In life, there is no second chance as you cannot take back the past. The story I tell is indeed one with many unexpected twists told in two parts that cover the past and the future. The history puts me at a time where my career spans the first generation of molecular biologists and the tsunami of data that now allows us to perform science in a way never before possible in human history. The second part recasts many of the early attempts to make sense of it all, using the insights gleaned from our information age where we are progressing from systems that run on transitive logic to those where intransitive designs allow systems to reset, repair, regenerate, replicate, reproduce, and reprogram themselves as they evolve.

Acknowledgements

I have only barely mentioned my family as they have their own stories to write. I am very proud of what my children have achieved.

Over these years, I owe my thanks to many colleagues along the way who helped during this journey of "job and finish" – Alex Rich definitely made it possible and Mike Christman at Boston University made it survivable. I thank many of my co-authors who were also curious and helped make it happen, especially Ky Lowenhaupt and Jeff Spitzner. Burghardt Wittig, with whom I never published, has also been someone I can turn to for support when I needed help.

Recent work with Sid Balachandran at Fox Chase and Maria Poptsova at HSE in Moscow and the junior scientists on their teams is exciting and ongoing. We are working on some really novel discoveries. I now also have a number of old and new friends with whom to collaborate on other flipon work. While many more fellow travelers are not mentioned by name, the details of our achievements can be found in the papers referenced at the end of the book.

I also am grateful for the editors who accepted papers in their journals way before it was certain that the job would be finished, especially the editor at *Communications Biology* and Caryn Navarro at Cell Press, who helped me back into the Z-DNA field. Ursula Weiss at *Nature* has helped greatly in recently advancing the Z-DNA and Z-RNA field. I also appreciate the help of Professor Steve Brown, Peter Robinson, and Professor Mike Shipston for their help with shepherding other manuscripts to press. The other failures along the way only reinforce the rubric "There is no gain without pain". Best of all, flipons have turned out to be a lot of fun!

About the Author

Dr. Alan Herbert's career spans both academia and industry. He is Head of Discovery at InsideOutBio and has published in high-profile journals. His work on left-handed Z-DNA and Z-RNA has proven that, in addition to the right-handed DNA double helix that Watson and Crick first described, higher-energy forms of DNA and RNA are used in the cell to regulate many important biological functions, including defences against viruses and cancer and to regulate the readout of information from the genome. His publications include one providing the first genetic evidence for the biological relevance of the Z-DNA/Z-RNA conformation. This work started with his discovery at MIT of the Zα family of proteins that recognize a form of left-handed DNA called Z-DNA. During this time, Dr. Herbert has had many fruitful collaborations with scientists worldwide who have made great contributions to the work. These discoveries led to the therapeutic targeting of flipons to treat solid tumors. In other endeavors, he helped pioneer the use of genome-wide association studies at the Framingham Heart Study and contributed to the development of new drug programs at Merck. Dr. Herbert has also helped to expand our understanding of the roles of the alternative DNA structures encoded by genetic elements called flipons during evolution. He has also detailed mechanisms by which the interactions between small noncoding RNAs and flipons have the potential to regulate the sequence-specific read out of genes as embryos develop. In the last few years, the work has changed our understanding of how normal cells function and how disease develops.

Part I

With the exception of our behavior, the past is not a good predictor of the future. Here, we examine why that is also true of science and scientists.

1 The Dawn

This book is a personal recollection of the events that established a role in biology for the left-handed conformation Z-DNA: a role so important that it changes our understanding of how our genome evolves and how we protect ourselves against viruses and cancers, while sparing normal cells. The insights cause us to rethink how information is encoded in the genome by DNA and RNA – by their shape as well as by their sequence. The discoveries were unexpected, even by the author.

This book is also written for those like the younger me, just to reassure these misfits that it will turn out OK, even though it may not always seem so along the way. My journey was quite an adventure, but also one with many moments of doubt about the wisdom of my choices. Here, I will answer a few of the questions that others have brought up along the way, such as "Who cares if you can twist right-handed Watson-Crick DNA the opposite way?", "Why does it matter that DNA can fold into different shapes?", "How does that cure cancer or explain anything of importance?", and "Why waste money on that kind of research when there are more pressing needs?"; even "How did you get fired from so many places?". The questions are similar to those asked of anyone who steps too far out of line. Well, this book is my response to those questions. The answers are all related. I will also address newer questions for which I don't yet have an answer.

WHAT IS DNA?

DNA is formally called deoxyribonucleic acid and it forms many different structures. The best-known conformation is right-handed and is called B-DNA while the left-handed helix is called Z-DNA (Figure 1.1).

The discovery that DNA was the genetic material took some time to gain acceptance. No lights went on when the experiments of Oswald Avery, Colin MacLeod, and Maclyn McCarty made any other explanation unlikely. Many thought of DNA as being too simple to perform the complex functions attributed to it. The DNA strands were only composed of four components, the nucleotide bases. Phoebus Levene, Oswald Avery's colleague at the Rockefeller Institute for Medical Research, said that DNA was just a long chain formed by repeating these four elements in the same order (Figure 1.2). The structure was therefore assembled the way carbohydrates are made from simple sugars and collagen is formed from proline repeats. Many scientists considered DNA to be only a scaffold to hang protein on, much as a clothesline is used to dry laundry. Furthermore, proteins were capable of performing many different biological functions. Therefore, proteins, and not DNA, carried the hereditary material. Why would Nature work anyway else?

In 1953, Watson and Crick [4], guided by the experimental findings of Maurice Wilkins [5] and Rosalind Franklin [6] (Figure 1.3), reasoned correctly that two anti-parallel DNA strands wound around each other to form a right-handed double

DOI: 10.1201/9781003463535-2

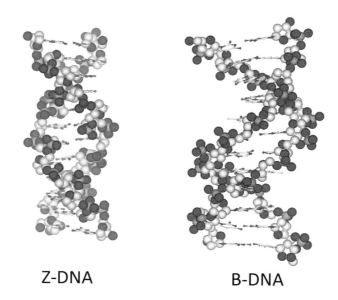

Z-DNA   B-DNA

FIGURE 1.1 The left- and right-handed conformations of DNA can switch from one form to the other by inverting the base pairs and do not require breakage of the helical phosphate backbone.

helix [7]. They proposed that each strand could be copied using complementary base pairing to produce two new molecules of DNA. The copying process would occur during cell division so that each daughter cell would receive a complete set of the DNA molecules present in the parent.

IN THE BEGINNING

Evidence that DNA was the hereditary material accumulated over time. The discovery in the early 1900s of bacterial viruses, called bacteriophages or phages, consisting of a single DNA strand coated in protein, was suggestive of this possibility. There was also no doubt that genetic traits in fruit flies traveled with the chromosomes. These tiny thread-like structures were transmitted from one cell to another and from one generation to the next. The threads were composed of DNA and protein and could be seen during cell division under a microscope (Figure 1.4A). That the threads were different from each other was revealed by the use of different dyes which showed that each chromosome had a distinct banding pattern. When cells divided, the number of chromosomes doubled, with one set of each chromosome sent to each of the two daughter cells

In addition to the banding patterns, chromosomes could be distinguished from each other by a number of other features, including variations in length or a distinctive feature, as was used to track the white eye gene in fruit flies to the Y chromosome (Figure 1.4B). That made it possible to show, using breeding experiments, the association of a specific trait with a specific chromosome. These findings left no doubt that chromosomes contained the hereditary material.

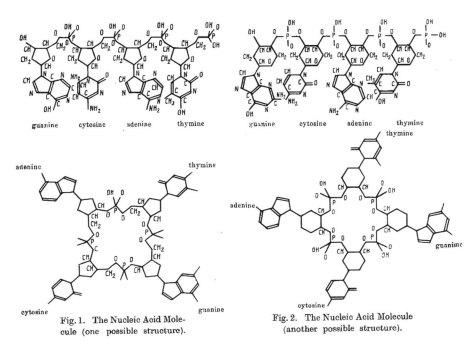

Fig. 1. The Nucleic Acid Molecule (one possible structure).

Fig. 2. The Nucleic Acid Molecule (another possible structure).

FIGURE 1.2 Possible nucleotide structure of DNA. From Wrinch, D.M. On the molecular structure of chromosomes, *Protoplasma* 25, 550–569 (1936).

The proteins in each chromosome were initially assumed to transmit the genetic instructions to each new cell. The proteins were a very diverse bunch. Many were enzymes able to catalyze the reactions required to build the components of a cell. Others facilitated the reactions that were needed to make a cell work. Collectively, proteins could convey all the information present in the old cell to the new cell. The

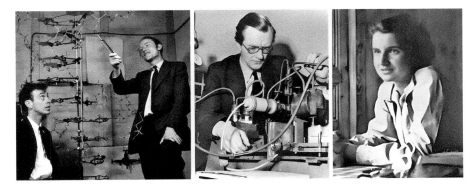

FIGURE 1.3 Jim Watson and Francis Crick next to their wire model of DNA, and Maurice Wilkins, a New Zealand expatriate like me, who, along with Rosalind Franklin, produced the experimental proof on which the model was based and refined.

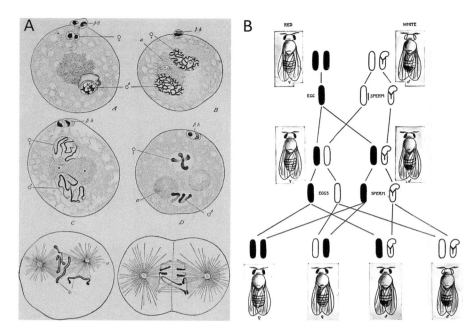

FIGURE 1.4   Early evidence of a role for DNA in the transmission of information from one generation to the next. A. From *The Cell in Development and Inheritance* (1902) by Theodore Boveri. B. From *A Critique of the Theory of Evolution* (1916) by Thomas Hunt Morgan.

search was then on for proteins able to join one type of amino acid specifically to another type. Those proteins would allow the production of new protein copies in a cell. At a minimum, 400 such proteins would be necessary to allow each of the 20 different amino acids to be coupled to any other amino acid. An existing template would ensure that the amino acids were added in the correct order. Most likely, the mold was another protein.

At the time, this proposal made sense. The role of DNA was disparaged for a number of reasons. DNA was built from just four different nucleotide bases as opposed to the 20 common amino acids that proteins used. Furthermore, Levene's idea, that the nucleotides in DNA were always linked in the same order, gained sway through his dominance in the field [8]. Another colleague at Rockefeller, Wendell Stanley, crystallized the *Tobacco mosaic virus* in 1935, a Nobel-Prize-winning feat, and claimed that the protein coat, or capsid, did not contain nucleic acid (see [9]). His findings convinced many that protein was the hereditary material. The idea was that the viral proteins provided a template for the cell to make more viral proteins. Much later, work revealed that the virus was 6% RNA by weight.

Other work performed at Rockefeller by Avery, MacLeod, and McCarty (Figure 1.5), however, provided irrefutable evidence that DNA was the staff of life, despite the claims made by others [10]. With DNA alone, McCarty and Avery could convert a harmless strain of pneumococcus to a virulently lethal bacterial strain. They

FIGURE 1.5 Oswald Avery, Maclyn McCarty, and Colin MacLeod.

could see the bacterial capsule change from rough to smooth during the transformation. Their Rockefeller colleagues, however, cast doubt on these results for one reason or another. Levene and Stanley were persuasive, but wrong. What was initially their "working hypothesis" became "entrenched" (p. 74 [11]) and a test of fealty.

The Watson-Crick model of B-DNA was not an obvious discovery, and is a story well told by Watson in his book, *The Double Helix*. The B-DNA Watson and Crick described looked like a helical staircase with the phosphate backbones of each strand acting as the handrails. One strand pointed up the staircase and the other downward (Figure 1.6). The helical steps were composed of bases known as adenine (A), thymine (T), guanine (G), and cytosine (C). "A" paired with "T" and "G" with "C", with such base pairs connecting the two strands. The model supported Chargaff's rule that the number of purine (A + G) present was equal to the number of pyrimidines (C + T). The pairing between purines and pyrimidines formed usingthe base geometries Jerry Donohue told Watson about.

The proposed structure provided a perfectly good description of how genes work and how hereditary material is transmitted from one generation to the next. Each strand was a perfect template by which to produce an exact replica of the other strand. Once each strand was copied and paired with its complementary strand, a cell could split in two with each descendant receiving an identical replica of its parent's genome, albeit with a few errors here and there. The success of the Watson-Crick model was magnified by subsequent work revealing how the order of DNA bases exactly specified the order of amino acids in proteins. The variations in DNA base sequence accounted for the different proteins made. The letters of the genetic code were the keys to decoding this three base cipher.

The genetic code is nearly universal, the same for almost all forms of life here on earth. That finding all by itself provides evidence for the evolution of all organisms from a common ancestor, as Darwin and Wallace imagined. The DNA has a dual

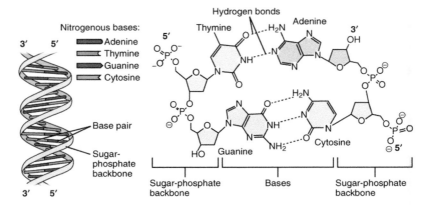

FIGURE 1.6 DNA base pairing. A cartoon of a right-handed B-DNA helix. The strands run in opposite, anti-parallel directions. The base pairing occurs between guanine and cytosine bases (G:C) and between adenine and thymine bases (A:T). Nucleotides consist of both the base and the sugar. The bases point away from the sugars and are in the anti-conformation arrangement (source https://commons.wikimedia.org/wiki/File:DNA_Nucleotides.jpg).

function. It provides the instructions for copying and transmitting itself and also includes the information necessary for the cells to survive until they produce their next generation. Those instructions are read out into a different nucleic acid polymer called RNA. DNA was the hereditary material, while RNA provided the plans for the cellular machinery.

DNA provided the template for both itself and RNA. The production of RNA did not necessitate the replication of DNA, allowing the two processes to occur independently of each other. The rate of RNA synthesis did not rely on when and how a cell divided. Information flowed from DNA to RNA to protein. This design allowed the amount of RNA produced by a cell to vary with the environment, while DNA replication could follow a different schedule. Due to the random assortment of chromosomes transmitted to progeny from parents, there was always a chance that the offspring were better adapted to the environment than their parents. Variations in how each cell produced RNA also meant that some cells were better adapted than other cells. While every cell came from a cell, no two cells ended up exactly the same.

The directions to make the protein on ribosomes were carried from the DNA in the nucleus to the cytoplasm by messenger RNA (mRNA). Acting on these instructions, the ribosome would then add amino acids to the nascent peptide chain in the correct order by catalyzing the formation of peptide bonds between the amino acids, building a protein step-by-step. A small adaptor, called transfer RNA (tRNA), contained an anticodon that matched the codon in the messenger RNA that was linked to only one of the twenty amino acids. The correct pairing of codon and anticodon would ensure insertion of the exact amino acid specified by the mRNA. The system required only a small number of proteins to check the match between tRNA and mRNA was correct to certify that the proper amino acid was added to the growing peptide chain. The design allowed a generic ribosome to make any protein, regardless of the amino acids involved in the synthesis.

The race was then on to find out how the DNA code was implemented. As George Gamow predicted, the code was a triplet, with a sequence of three bases specifying a particular amino acid (Figure 1.7) [12]. Some triplets were reserved to tell the ribosomal machinery where to start and stop the translation of RNA into protein. The RNA triplets were found not to overlap, but there was more than one codon coding for each amino acid (Figure 1.7). Interestingly, there are 64 possible triplets ($4 \times 4 \times 4$), but only 51 are found in Nature (Figure 1.7).

The discovery of the genetic code exemplifies the explanatory powers of science where simple principles lead to the best-available description of complex outcomes. These findings, once understood, provided a roadmap for the development of the biotechnology industry, enabling the delivery of innovative products and therapeutics to improve health. With such advances in hand, was there a need to look for more? Why would you? The discoveries delivered beyond whatever was expected! All such discoveries have their Charles Duell, who claimed in 1899 that there was nothing more to invent (see http://tinyurl.com/3ttb66jr), or their Lord Kelvin, who dogmatically attested that "Heavier-than-air flying machines are impossible" (although no source is ever given for this quote).

Second Position of Codon

		T		C		A		G		
	T	TTT / TTC	Phenylalanine F / F	TCT / TCC	Serine S	TAT / TAC	Tyrosine Y	TGT / TGC	Cystine C	T / C
		TTA / TTG	Leucine L	TCA / TCG		TAA / TAG	STOP end / end	TGA / TGG	STOP end / Tryptophan W	A / G
F I R S T	C	CTT / CTC / CTA / CTG	Leucine L	CCT / CCC / CCA / CCG	Proline P	CAT / CAC	Histidine H	CGT / CGC / CGA / CGG	Arginine R	T / C / A / G
						CAA / CAG	Glutamine Q			
P O S I T I O N	A	ATT / ATC / ATA	isoleucine I	ACT / ACC / ACA	Threonine T	AAT / AAC	Asparagine N	AGT / AGC	Serine S	T / C / A
		ATG	Methione START M	ACG		AAA / AAG	Lysine K	AGA / AGG	Arginine R	
	G	GTT / GTC / GTA / GTG	Valine V	GCT / GCC / GCA / GCG	Alanine A	GAT / GAC	Aspartate D	GGT / GGC / GGA / GGG	Glycine G	T / C / A / G
						GAA / GAG	Glutamate E			

FIGURE 1.7 The genetic code. The triplet codes specify the amino acid inserted into a protein. They also specify where to start and stop translation. As we will discuss later, RNA messages can be recoded by adenosine-to-inosine editing. Those edits that change the amino acid inserted into a protein are highlighted in dark gray, while those edits that do not change amino acid coding are in light gray. Editing of the ATA isoleucine codon causes insertion of either methionine or valine.

Following these discoveries, further experiments focused on bacteria as they were the easiest to work with. Bacteria divided fast and the cookbook of recipes made experiments easy. Their genomes were lean and efficient. Each gene was no bigger, nor any smaller than it needed to be, honed by evolution to do the most with the least. Their ordered array of genes allowed for a rapid and efficient response to a change in the environment. Their linear genomes enabled the mapping of genes for biochemical pathways. The way functionally related genes were clustered led to the concept of operons that allowed expression of all the key genes in a pathway at once. Further studies provided insights into how operons were regulated. Expression of the genes was turned on by metabolic precursors of the pathway and switched off when not required. Combined with the much earlier discoveries of Koch, Pasteur, and their colleagues, the studies of fungi, especially those already industrialized for food preservation through the ages, helped validate and extend these concepts.

These findings were made at the time computers were coming into their own. Building on these advances, Norbert Wiener established the field of cybernetics that focused on control circuits to regulate machines, brains, or whatever process held your interest [13]. The operon concept was a perfect fit [14]. Life was just cybernetics cloaked in biological garb, with nanoscale wiring diagrams recorded in DNA; the new insights led to an understanding of cells as machines that could eventually be engineered.

Was it time to declare mission accomplished? Well … no. It soon became apparent that the paradigm of DNA to RNA to protein was not quite the entire story. The simplification of life needed revision. Sure, there were RNA viruses that never felt the need to encode their genome in DNA, where RNA acted as the template for both replication and translation (Figure 1.8). The viruses spread by having the infected cells make as many RNA copies of both template and message for them. This required just a minor amendment to the dogma as RNA viruses were still bound by the same genetic code as their hosts. Their reliance on the host machinery for their replication gave them no other choice. Surprisingly, some RNA viruses were found to copy themselves into DNA. That exception was easily accommodated into

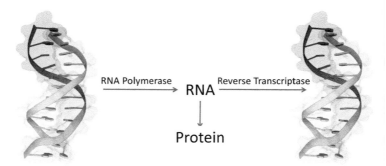

RNA Polymerase   RNA   Reverse Transcriptase

Protein

FIGURE 1.8 RNA can be used as the template for both DNA and protein (adapted from https://commons.wikimedia.org/wiki/File:DNA_Orbit_Animated_Clean.gif).

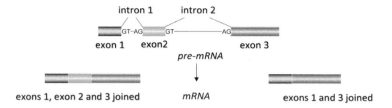

FIGURE 1.9   RNA splicing. The introns are removed so that the exons join to make the message that instructs the ribosome on how to make protein. The GT and AC indicated the dinucleotide repeats that indicate the beginning and end of an intron.

the grand scheme once the protein responsible for synthesizing DNA, using RNA as a template, was discovered [15, 16].

Eventually came the discovery that viruses weren't so simple after all. Another revision to the plan was required. Some viruses could encode multiple proteins in their genome by splicing pieces of RNA together (Figure 1.9) [17, 18]. This discovery also came as a surprise. Yet, the realization that the initial RNA transcript from DNA was not the final message sent to the cytoplasm occurred almost simultaneously to a number of investigators, though many made the connection only with hindsight. They had the data, but not the correct explanation. The possibility that a gene consists of DNA pieces separated by vast stretches of irrelevant sequences did not occur to them, even at their smartest moments. Why would you ever believe that "higher beings" like humans were so much less efficient at making RNA than evolutionary reprobates like bacteria? Why would the human genome be defective by design, delivering RNA in a state unable to produce a proper protein? Why would you require a reassembly of the initial RNA transcript by the cellular machinery to decode the message sent from the genome to the ribosome?

It seemed wasteful to expend so much effort to correct the needless errors present in the what came to be known as pre-mRNA. Processing required removal of RNA that did not code for protein (called introns) from the transcript and then splicing together the exons to make the correct mRNA to make the desired protein. It all seemed so wasteful. Much energy was spent just to make introns, and then more was required to eliminate them. There was also the problem of where to cut the RNA and join the exons correctly. Clearly, there had to be a motif or code of some kind or another. To this day, no one knows exactly the rules used by the cell to process the RNA correctly. We know that the retained sequence is bracketed by two particular bases at either end of the exon (Figure 1.9; GT at one end and AC at the other). However, each pair of bases would be expected to occur by chance once every sixteen nucleotides, so the insight is not particularly informative. That the strategy works is self-evident, but not so evident as to reveal the secret of how the cell achieves such success. The only thing that is certain is that the laws of thermodynamics apply and much of the energy required is expended as entropy.

In the final analysis of the human genome, only about 2.6% of the DNA codes for proteins. On a traditional grading scale, that success rate is an automatic fail.

The outcome is nowhere close to a perfect score. If you include introns in the total, genes represent up to 40% of all sequences; still, much of the transcript is made and never used. The other 50% or so ... what's that all about? Is it just "junk" (a description attributed to Susumu Ohno (see [19], pp. 366–370)? In total, around 54% of the genome is repetitive, consisting of large families derived from short blocks of similar sequences. Sure, some of these sequence blocks prevent the ends of chromosomes from fusing with each other during cell division and so protect the genome. Other repeat sequences seem to ensure that the correct set of chromosomes is passed to each descendent by allowing each copy to align and check the other one out. Pairs of chromosomes may kiss, but the attachment is fleeting. But does this "junk" really do anything useful in a cell? We will see. One simple answer is that splicing increases the diversity of RNAs that can be made from a gene – it allows assembly of exons into different combinations to generate a more diverse set of proteins. Each mRNA extracts a different set of information from the genome to create a unique ribotype, allowing a cell to express its own particular phenotypic personality (Figure 1.10) [20].

The idea of junk DNA was an affront to geneticists like the opprobrious Sidney Brenner (see [21]). The extra letters just made genome sequencing more complicated, costly, and a lot more work. Furthermore, these repetitive sequences were often so similar that it was hard to know where to place a particular sequence in the genome assembly. Better to find out the important stuff by sequencing the puffer fish, whose genome was lean in comparison to the overstuffed human genome [22]. Clearly, the puffer fish did not need all those extraneous letters to survive. Yet, as we will find out, all that "junk" in the human genome increases its capacity to encode a complex catalog of choices.

However, the money play was in the human sequence. There was gold in those letters that could be mined by pharmaceutical companies to make many more billion-dollar molecules. The race was on! Although the impression was given that the human book of life was snatched from the gods in 2001 by the modern-day Prometheans, Craig Venter and Francis Collins, it was only in 2023 that the job of end-to-end sequencing of every chromosome was completed by the many thousands of scientists who really did the work. Then, that 2023 sequence was only for a single

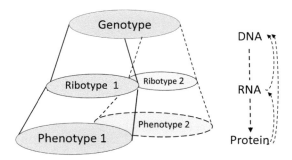

FIGURE 1.10   Different sets of information can be read from the genome to create ribotypes that specify different cellular phenotypes.

genome from a single cell. We already know that the human genome comes with many different alternatives. The option you received reflects on your ancestors' success in conquering a particular realm of the biosphere. But is that really the whole story? Will more sequencing provide the Rosetta Stone, enabling us to decipher the code of life even further? Is there a hidden logic in our genome that we know nothing about? Does our DNA embed a code for immortality? Does the "junk" contain other types of secret messages? You will never find out that from puffer fish DNA, especially if you insist on eating this tetrodotoxin-laden fish as a delicacy. Instead, the only thing you will likely learn about through your interactions with the pufferfish is your own mortality.

All we can say is that the genome is not designed according to the sound engineering principles imagined when the genetic code was cracked. As George Gamow showed, Nature is clearly not a mathematical genius: there are more efficient and elegant ways of encoding information than are used in biology. In fact, Nature does not care how clever the code is, only that the code works. Indeed, Nature's fight is with entropy. Any information must first be discernible above the noise. When Nature reads the genetic code, it does so mostly by Braille, as it does not browse each letter individually. Rather, it feels for the specific shape produced when a tRNA anticodon is an exact match for the codon present in the mRNA message. If the shape is wrong, then there is an error in the decoding. If the shape is right, then it is safe to proceed with translation. So, maybe we should not expect Nature to work in ways that smart human beings can conjure up. Maybe the junk is full of surprises and does something useful? Maybe the repeats are more than just misshapen stutters in the language of life? Does life beget such sloppiness or does such sloppiness beget life? Or maybe the junk is just there for no reason at all? We will address all these questions in later chapters.

2 Even Smart People Are Sometimes Wrong

This book is about scientific discovery. About working on an idea that was premature. About a comeback for a field down for the count. How could the smartest people in the world be so wrong? (They were.) Shouldn't they know better... after all, they are smart, aren't they ... with IQs well above 100? This book is about an unusual form of DNA, called Z-DNA. The uniqueness of Z-DNA is that it is a left-handed double helix; the two strands in DNA wind around each other by twisting to the left. The famous Watson-Crick, which encodes genetic information by the sequence of bases in the helix, is called B-DNA and twists to the right. The bases on each strand pair with those on the other strand according to specific rules. The base pairs then connect the two rails of the helical staircase. Z-DNA encodes information in a different way, as we shall see.

The discovery of Z-DNA was unexpected. The announcement came with banner headlines. Found by accident, Z-DNA was not part of any story. It was an answer to a question no one was asking (Figure 2.1). At the time, everyone wanted to crystallize DNA and determine in atomic detail the exact structure of the Watson-Crick DNA. The possibility of making DNA crystals had just arisen due to the recent advances in the chemical synthesis of DNA's building blocks that for the first time in history allowed the production of any DNA sequence so desired.

By 1979, the Watson-Crick model had aced a number of tests, but had not been fully confirmed by experiment. Certainly, the long, mucous-like DNA fibers drawn out with a glass rod from concentrated DNA solutions had been studied with X-rays. The hypothesized B-DNA model was the best current fit for the best available X-ray diffraction patterns. The simplest test for the model had been to build a scaled-up version and show that the proposed structure was reasonable given generally accepted chemical principles and the known properties of DNA. Then, under the glare of a bright monochromatic light source, the model would be placed at one end of a long, wide corridor and imaged on the wall at the other end. If the model was correct, the pattern visualized on the wall would match that recorded from the DNA fibers using X-rays. The invisible X-ray source, with its angstrom wavelength, was thus replaced by visible light with a nanometer wavelength. Everyone could see with their own eyes whether there was an acceptable match between the X-ray photographs and the images at the end of the corridor (Figure 2.2). Then, the tedious process of exactly calculating the fit by hand began. The Watson-Crick proposal gave a good fit to the data. The model was not perfect. Maurice Wilkins, who shared the Nobel Prize with the two gallants, subsequently spent many years on the further refinement of the structure.

DOI: 10.1201/9781003463535-3

Z-DNA Moves Toward "Real Biology"

Once considered to be an oddity of no particular significance, this unusual structure is now showing up in DNA segments that control gene expression

Gina Kolata Science Vol. 222, No. 4623, Nov. 4, 1983, pp. 495-496

FIGURE 2.1   So if it's in *Science* magazine, then it must be true, right???

However, the issue was not completely decided as another model with the two DNA strands arranged side-by-side, rather than intertwined, was still considered possible (to this day, some still think this is the correct model). After all, wouldn't a side-by-side arrangement make more sense? Separating the two DNA strands after completing their replication was just a matter of pulling them apart. There was no need to break the strands presented in a helix like the one Watson and Crick proposed. Further, did Watson and Crick have the contacts between the base pairs correct? No one knew for sure. Obtaining a crystal and determining the structure at atomic resolution was the best way to finally confirm the Watson-Crick DNA model. Once DNA could be synthesized on a machine, the race began to actually crystallize DNA and be the first one to solve the structure.

Two groups made crystals of the same small DNA sequence based on alternating d(CG) repeats around the same time. Horace Drew and Richard Dickerson at Caltech had d(CG)$_2$ crystalized in 1978 [23] but then Andy Wang, Gary Quigley, and Alex Rich at MIT solved the structure of d(CG)$_3$ first in 1979 [24]. Jacques von Boom synthesized DNA for the MIT team. The choice of d(CG)$_3$ was based on two things – the fact that the sequence would bind to itself as the G on one strand would base pair with the C on the other strand (Figure 1.6). According to the Watson-Crick DNA model, the synthetic DNA would naturally pair with itself in an anti-parallel orientation to form the double helix. Secondly, the pairing would most easily happen with G:C base pairs as they were more stable than A:T base pairs since they have three hydrogen bonds between them, rather than just two (Figure 1.6).

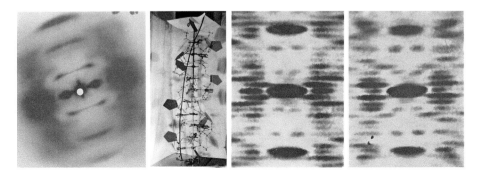

FIGURE 2.2   The X-ray photograph of rat tail collagen compared with the deduced wire model and the optical map for the collagen I and collagen II structures (from Watson and Crick. *Journal of Molecular Biology*, 3(5), 483–506, 1961).

The DNA crystallized easily and diffracted X-rays remarkably well. The images obtained were highly detailed. There was just one problem. The X-ray pattern did not fit the Watson-Crick model of B-DNA! Even though atomic resolution data were produced, meaning you could easily visualize individual atoms, something was definitely wrong! (Figure 2.3)

At this point, every scientist looks for the obvious mistake – was something wrong with the DNA synthesis, with the diffraction equipment, or with the X-ray collection? Did the crystal actually contain DNA? But there was nothing wrong with any of those things. The X-ray diffraction patterns gave 0.9 Å resolution (that is 10^{-10} of a meter). You could clearly see the cloud of electrons that form as a ring over the nucleotide bases. You could even see a hole in the middle of the ring where there are no atoms! Finally, the structure was solved. The base pairing between the two DNA strands proposed by Watson and Crick was correct. But no, it was not the Watson-Crick DNA.

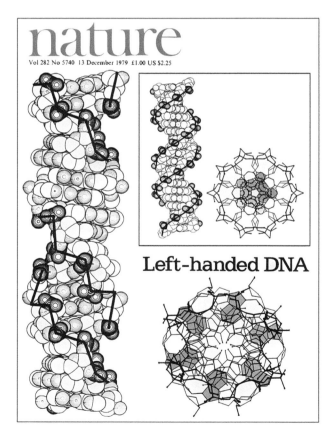

FIGURE 2.3   The first crystal structure of DNA was a left-handed double helix, not right-handed as expected from the Watson-Crick structure

Left-handed Z-DNA   Right-handed B-DNA

FIGURE 2.4   Left-handed Z-DNA *versus* right-handed B-DNA

What was wrong? The DNA twisted the wrong way! The helical backbone was left-handed, not right-handed as expected. Not only that, but the backbone was not just a mirror image of right-handed DNA. The backbone was kinked, rather than smooth. The kinks repeated every two base pairs. The new helix was called Z-DNA because the backbone zigged and zagged (Figure 2.4).

The name placed the helix at the opposite end of the alphabet to B-DNA and to the right-handed, double-stranded A-RNA helix. The bases were upside down relative to B-DNA, suggesting that forming Z-DNA only required an inversion of the base pairs. There was no need to break the DNA backbone. The kinks were due to the two distinct nucleotide conformations in Z-DNA. In B-DNA all the bases point away from the sugar that connects them to the phosphate backbone (the *anti* conformation, Figure 1.6). In Z-DNA, the dG base actually bent back to lie over the sugar ring (the *syn* conformation). The change in conformation occurred for even two residues, producing the zig-zag effect.

What a surprise! It was as if Cinderella was invited to the ball, but the fairy godmother had a sense of humor. She did not send Cinderella, but instead a maiden of equally stunning beauty. The purpose was to test whether the prince was fickle with his affections. Eventually, Cinderella turned up in a crystal carriage composed of a different DNA sequence. We all know how that story turned out. The Prince and his Court only wanted Cinderella, not the rather striking, but odd-looking sibling.

The hype surrounding the initial discovery of this left-handed Z-DNA was itself unexpected and spectacular. There were many roles proposed (Figure 1.9). As Dr Suess might write "Why, the things that Z-DNA can do". The outpouring was the biological equivalent of the theory of everything. There was definitely a lot of premature speculation. Journalists were covering the story as if they were paparazzi tracking the illegitimate child of a noble lineage – well, not really. Although DNA was involved in both cases, the story was not tabloid fodder.

THE FIRST SIGHTING OF LEFT-HANDED Z-DNA.

Surprisingly, earlier work had provided evidence that left-handed DNA existed. The evidence was indirect. Fritz Pohl was fascinated by the chirality of biological molecules (where chirality refers to whether something is left- or right-handed). Fitz first suggested, on the basis of spectroscopic studies, that a left-handed helix was formed when DNA was placed in a high salt solution (6 M sodium perchlorate whereas 150 mM NaCl is physiological, i.e., a concentration 40 times lower than the physiological one). The interpretation was supported by using circularly polarized light to confirm the difference in chirality between regular DNA and the high-salt form. Joined by Tom Jovin, the pair investigated the finding further using long polymers of synthetic DNA. They showed that alternating adenine and thymine polymers did not undergo the transition, but those with alternating guanine and cytosine did. Interestingly, alternating inosine with cytosine resulted in a different right-handed spectrum but it did not undergo a salt-induced transformation to produce a left-handed DNA spectrum. The result contradicted an earlier *Nature* publication claiming that this polymer formed left-handed DNA, a claim that is not correct. Every time I say that Z-DNA was a surprise when the crystal structure was solved, Tom Jovin tells me that he and Fritz Pohl were two people who were not surprised by the discovery of left-handed DNA [25]. They had been saying for years that such a conformation existed. Tom was the one who encouraged Horace "Red" Drew to try and obtain a d(CG)$_2$ crystal. The crystal Red obtained was from high-salt conditions, had a slightly different orientation of the phosphate in the backbone, and was called Z' compared to the MIT low-salt crystal.

So, was the conformation of the d(CG)$_n$ polymer in high salt found by Tom and Fritz really the same as that found in the crystal? The answer was yes. With the crystal in hand, a variety of spectral techniques confirmed that both the backbone and sugar resonances matched the alternative polymer structure. That was reassuring but the high salt concentration used to flip the polymer was nowhere near what is found inside cells. Interestingly, the right-to-left transition will occur under physiological salt conditions in the presence of metal ions when the d(CG)$_n$ polymer has a modified cytosine with a methyl group at the 5-position of the base (5-methylcytosine). The positively charged metals reduce the repulsion arising in Z-DNA from the closely approximated and highly negatively charged phosphate groups in the backbone. In retrospect, Tom and Fritz could have used this same polymer in their studies to obtain the low salt result: it was an intermediate in their synthesis of poly-d(G-C), but they did not test that particular DNA polymer.

The finding that Z-DNA might form under physiological conditions generated a new wave of excitement as 5-methylcytosine occurs in natural DNA. A further observation by the Hingerty lab provided another route towards stabilizing Z-DNA under physiological salt conditions. They found that the mutagen N-acetoxy-N-2-acetylaminofluorene (N-acetoxy-AAF) stabilizes dG in the *syn* conformation [26]. Subsequent studies showed that increasing amounts of the N-acetoxy-AAF adduct stabilized Z-DNA in the presence of increasing amounts of alcohol. Indeed, the C8 of dG, that was targeted by N-acetoxy-AAF, was shielded by the phosphate chain after the flip to Z-DNA but exposed to solvent in B-DNA. When this same C8 residue was brominated by Achim Moller in Alex's lab to modify about one-third of the dG, the polymer adopted the Z-DNA conformation under physiological conditions without added alcohol [27]. The C8 modification does not occur naturally but proved useful for making antibodies against Z-DNA. The antibodies, made by Eileen Lafer and Dave Stollar at Tufts, turned out to be very valuable reagents [28]. The finding of natural Z-DNA antibodies in patients with systemic lupus erythematosus also provided a marker for disease activity and set the scene for later studies by David Pisetsky at Duke in patients with autoimmune disease.

The big leap forward was showing that Z-DNA could be stabilized in plasmids. These are circles of double-stranded DNA made by joining the two ends of linear DNA (Figure 2.5). The result was first shown by the Wells lab using high-salt gels to induce the B- to Z-DNA flip [29]. The change in helical twist from right to left was compensated for by a change in the number of times the two strands writhe around each other. The two opposites cancel each other out without any need to break the DNA backbone. What this means is that a plasmid with a segment of Z-DNA has less writhe than when the segment is in the B-DNA conformation, as shown in Figure 2.5. The sum of the twist and writhe together, referred to as the linking number of the plasmid, remains constant. What changes is the proportion of each in the plasmid. The final ratio represents a minimal energy state for the plasmid under the conditions studied.

It is possible to change the linking number of a plasmid by using enzymes called topoisomerases that cut and religate the DNA strands but keep the plasmid circular.

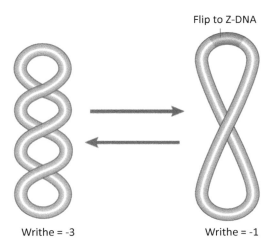

Flip to Z-DNA

Writhe = -3   Writhe = -1

FIGURE 2.5 When a segment in a negatively supercoiled plasmid flips to Z-DNA, the twist of the DNA segment changes from right to left. In closed circular DNA, there is a compensatory change in writhe τ (i.e., the number of times a plasmid wraps around itself) as seen by comparing the plasmid of the left to the one on the right (from *Nat. Rev. Genet.* 4, 566–572 (2003).

The same plasmid but with a different linking number is called a topoisomer. The plasmids have the same DNA sequence but differ only in their topology. The difference in linking number between two topoisomers is usually expressed as a difference in their supercoiling. A relaxed plasmid with no writhe and only B-DNA conformation has zero supercoiling. Supercoiling is negative when the twist of the plasmid DNA is less than is expected for the relaxed plasmid, causing the plasmid to writhe. The movement of negatively supercoiled topoisomers through the pores of an agarose gel depends on their writhe, which reflects the altered twist in the plasmid DNA.

Soon, the Wells and Jim Wang labs showed that the B-to-Z DNA transition could occur under physiological salt conditions by increasing the negative supercoiling of a plasmid. They could force the transition from B- to Z-DNA. Alfred Nordheim led the charge by the Rich lab to extend this work, bicycling to Harvard to coordinate the work with the Jim Wang team. Using two-dimensional gels, Larry Peck in Jim Wang's lab could separate topoisomers so that they could be seen with the naked eye (Figure 2.6) [30]. Further more, you can see when there was sufficient negative supercoiling to induce the transition from B-DNA to Z-DNA. The sudden change in twist caused by the flip to Z-DNA produced a decrease in writhe. The topoisomer then moved more slowly than an equivalent topoisomer without the Z-DNA-forming

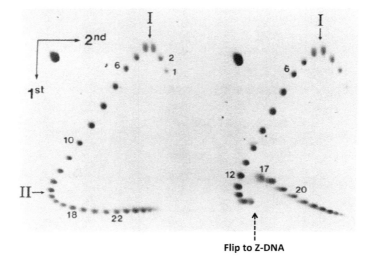

Flip to Z-DNA

FIGURE 2.6 Flipping DNA from a right-handed to a left-handed conformation alters the movement of a piece of circular, double-stranded DNA that is called a plasmid. Here, two different plasmids are present in the same gel. Each spot represents a variant of the plasmid, called a topoisomer, that is underwound by one DNA turn relative to the spots above and below it. The dot at the upper left of the gel is a relaxed plasmid with one or more nicks in the DNA strands. The gel is run in the downward direction to separate the topoisomers, then in the left-to-right direction under different conditions to better visualize the Z-DNA flip. The plasmid on the left does not have a Z-DNA forming sequence (from Peck and Wang *Proc. Nati. Acad. Sci. USA*. 80 (20) 6206-6210, 1983, permission from Jim Wang).

insert. The point of transition from B-DNA to Z-DNA is highlighted by the arrow in Figure 2.6.

But where did the supercoiling come from under normal conditions? The answer was supplied by Leroy Liu, also working with Jim Wang. What Liu and Jim Wang realized was that negative supercoiling could occur behind an RNA polymerase due to the untwisting of the DNA strands during transcription. The negative supercoiling was sufficient to power Z-DNA formation by a subset of sequences under physiological conditions. These sequences were most often the d(CG) or d(GT) dinucleotide repeats seen in crystals that adopted the Z-DNA conformation most easily. As more genes were sequenced, it became apparent that these sequences were in gene promoters. That's it, everyone agreed! Z-DNA regulates gene expression!

When I applied to move to MIT in 1985, Z-DNA was exotic and exciting. The headline for the article written by Gina Kolata of Science magazine at the top of this chapter said it all: "Z-DNA Moves Towards 'Real Biology'" (Figure 2.1). The work summarized by Gina indicated that Z-DNA binding proteins had been isolated, Z-DNA could be stabilized under physiological conditions by negative supercoiling, and that Z-DNA was detectable in chromosomes by using Z-DNA specific antibodies to stain them. Why wouldn't someone want to work on that and learn molecular biology at the same time? Alex wanted me to apply for a National Institutes of Health (NIH) Fogarty Fellowship to fund my stay in his lab. That done and successfully awarded, the time was set for me and my family to leave New Zealand and move to Boston.

3 Coming to America

> Z-DNA is an example of a discovery made by accident, where, however, belief in serendipity has so far led those who adopted it to a dead end.
>
> **Michel Morange [31]**

I arrived in Boston in August 1985. What a difference 24 months made! Jean L. Marx was about to publish, in the November issue of *Science*, an update to the 1983 article by Gina Kolata (Figure 3.1). It was a piece based along the lines of "The Emperor has no clothes". Alfred Nordheim, featured in the earlier piece, is quoted as saying ""What is important to realize, we don't have direct evidence for the *in vivo* existence of Z-DNA or for its physiological function". Furthermore, he said that "...the chromosomal staining by Z-DNA antibodies may just be due to the methods used to prepare the chromosome". He knew something was amiss and had already packed his bags for his return to Germany. His timing was perfect. The article noted that the "...best evidence so far in support of a physiological role for Z-DNA comes from William Holloman of the Cornell University Medical College in New York City and Eric Kmiec of the University of Rochester". Not really; it seemed that no one was able to reproduce Dr Kmiec's results, nor were they able to reproduce work he did subsequently in the Worcel lab nor his later work on chimeraplasty (http://www.lobbywatch.org/archive2.asp?arcid=1142). Somehow, the talented Dr. Kmiec managed to remain funded and keep his academic job. Others who tried to reproduce his results were not so lucky.

The role of Z-DNA in the control of viral gene expression was also being called into question. Results from the Herr lab at Cold Spring Harbor in New York and the Chambon lab in Strasbourg, France did not yield clear-cut answers. Mutation of the proposed Z-DNA-forming sequences in the SV40 viral regulatory region decreased gene expression but so did mutation of other nearby sequences that were not Z-DNA-forming. Notably, viral mutations that rescued replication after mutating the proposed Z-DNA-forming sequence did not restore a Z-DNA-forming sequence. However, the regulatory region was quite complex and different elements appeared important in different conditions. In addition, Patashne, Peck, and Wang from Harvard argued against Z-DNA formation having anything to do with gene expression in bacteria. The critics were quite vocal and sang the same tune. No punches were pulled. Alex stood alone in his corner of the ring. The decline in publications on Z-DNA told the story (Figure 3.2).

My first personal experience of this was when I went to a talk Jim Watson was giving at Harvard. Of course, the room was packed. With a youthful enthusiasm, I ventured to ask a question at the end of the talk. I asked Jim "What do you think the function of Z-DNA is?". He drew a breath, gave his crooked smile, and said "You should go down the road and ask Alex Rich, he is the only one who thinks there is

DOI: 10.1201/9781003463535-4
This chapter has been made available under a CC-BY-NC-ND license.

Z-DNA: Still Searching for a Function

Six years after the discovery of Z-DNA questions remain about whether it exists naturally and what its functions might be

Jean L. Marx Science, Vol. 230, No. 4727, Nov. 15, 1985, pp. 794-796

FIGURE 3.1   So if it's in Science magazine, then it must be true, right???

PubMed Citations Per Year (Per 100,000)

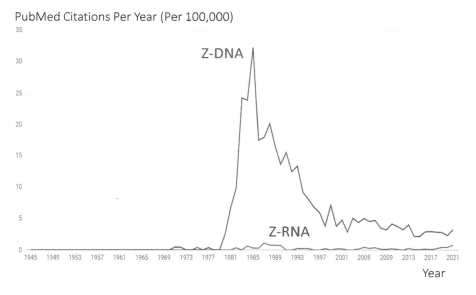

FIGURE 3.2   The hype about Z-DNA, then the Crash. The publications per year on Z-DNA and Z-RNA per 100,000 publications in the PubMed database. The Z-DNA crystal structure was obtained in 1979.

one". That brought the house down with hoots of laughter. As the Z-DNA frenzy waned, the circus tents were folded and the crowds moved on to the next event.

Others entranced by the mystery of Z-DNA had a similar experience. Nacho Tinoco commented in 2014, after he had established in 1984 that there was also a Z-RNA helix, "Left-handed Z-RNA provides another example of the attitude of some biologists to supposedly nonbiological results. Before a seminar I was going to give at a university, a professor walked in and said he looked forward to hearing about my work on Z-DNA. I said Z-RNA, and he turned around and walked out. After my talk on Z-RNA at the 1984 Gordon Conference on the Physics and Physical Chemistry of Biopolymers ... someone asked me what its biological relevance was. I responded that it was of interest to at least one biological organism: me".

Given the lack of success, the journalists sensed blood in the water and wrote more "must read" articles of the "Can you believe this" kind of exposé. The skeptics, who were in the majority, enjoyed the exposés. The critics then wrote a review of these reviews, couched, of course, in carefully crafted language, stopping just short of allegations. Z-DNA was panned as just another one of those oddities that turn up every once in a while, just another hype cycle that provided lessons to students about venturing down rabbit holes and led to a few jokes about Alex in Wonderland.

We can now say that everyone was wrong. Neither protagonists nor antagonists anticipated all the twists in the plot line – like any good story, there was hope at the start, then tales of hubris, and finally a surprise ending. But, first more of the history.

When I arrived in Boston, I was picked up by a friend, Tony Bierre, who had also been in the Pathology Training program in Auckland, but was now doing a Fellowship in Boston. He and his wife had just had a daughter and the apartment was small, so it was really kind of him to do that. It was lucky he was able to help as when I finally figured out how to get to MIT, find the right entrance, the right building, and the right lab, it seemed to be a surprise to everyone that I was there. Didn't I know how bad things were?

I learned that Alex would not usually appear until 3 pm and often stayed late but kept unusual hours to chat with colleagues in Europe and elsewhere. I was to learn that it was often difficult to separate Alex from his phone and that he was constantly adding to his palm-sized black notebook as he talked and made appointments in his calendar. His other favorite thing to do was to use a dictaphone to record notes or compose letters, punctuated by the odd air swallow and burp as he searched for words that weren't there. Or he would dictate a paragraph for his secretary and then say "don't type that". He seemed to be constantly editing the tape verbally the same way you and I would change text in a written paragraph. I don't believe he ever graduated to computers.

I also received a preview of the Jean La Marx article from Eileen Lafer and Mike Ellison, post-docs at the time. They too had joined the lab with high hopes but were definitely working on a fast exit. They laid out strategies to have a successful experience at the Rich lab, none of which included working on Z-DNA-binding proteins, given the questionable nature of the previous data. Eileen had observed anti-Z-DNA antibodies in a subset of patients with autoimmune disease and subsequently made a monoclonal antibody specific for Z-DNA. Making this antibody entailed immunizing mice with Z-DNA, then fusing antibody-producing B cells from those mice with immortalized cells. The hybrid cell produced had the desirable properties of both parent cells – the ability to produce an antibody highly specific for Z-DNA by a cell that could be maintained in cell culture forever or stored frozen for use when needed. She was using a similar technique to produce antibodies against proteins from bacteria that bound to a column containing Z-DNA. Mike was working on the energetics of the B-DNA to Z-DNA transition in plasmids, using different nucleotide sequences to determine the number of supercoils necessary to flip them from right-handed to left-handed helices. In two-dimensional gels, this caused a characteristic hump as the total writhe of the plasmid would be converted into the twist of DNA, causing its mobility in the gel to change as described above and shown in Figure 2.6. Mike was extremely cynical, perhaps the most cynical person I have ever met. While amusing at first, his humor can go to the dark side quite quickly. His survival guide was to work with the crystallographers or go elsewhere. He was not advocating a clean break. He was advocating what he termed Alex's cuckoo strategy. Here, Alex would embed his people in another laboratory, much like a cuckoo lays its eggs in another nest, to be incubated by its owner, later to claim the new offspring as its own. Apparently, that was the reason that Alex was associated with the cloning of interleukin-1. His postdoc Phil Aurin had done that in Lee Gehrke's lab. Alfred Nordheim's version was called the bicycle strategy because of the time he spent pedaling between the Rich and Wang labs. One look around the lab was enough to start me seriously looking for an alternative. It seemed to be the custom of departing people to just lay everything down as they had finished using them and then just walk out … or maybe it was a sprint once they saw a chance to successfully exit. The new people would then clear out a space to do their work and pile what was not needed in a corner or some flat surface, anywhere! Coming from the really nice labs I had experienced in Auckland, it was akin to doing science in a garbage dump, admittedly one of potential interest to archeologists specializing in the early history of molecular biology. The lunchroom was the most habitable space, which was probably why Mike made it his office. There was the cage in one corner for the pet iguana that had perished some time earlier, another relic of a lost soul. Literally and figuratively, the whole situation looked like a mess requiring a lot of cleaning up, not exactly what I had signed on for.

When I met with Alex later, I found his office was somewhat similar, with piles of papers scattered around the room, covering every elevated surface and slowly spreading over the floor. Instant thoughts of Bleak House and the risk of spontaneous combustion. We met only briefly, with many phone calls interrupting our conversation. Still, I had many things to keep me busy in the first weeks: obtaining the MIT IDs and health insurance, opening a bank account, obtaining a Massachusetts driver's license, buying a car, learning to drive this enormous Chevy V8 on the left-hand side of the road on Boston's narrow streets (in New Zealand, we drive on the other side in much smaller cars), and locating somewhere for the family to live once they arrived. I note that Alex's paper collection eventually turned out to be quite valuable. Apparently, there is a market among collectors for off-prints of famous scientific papers.

Chris Fredericks, who noted my immunology background and my desire to learn molecular biology, suggested that I might want to talk to Susumu Tonegawa. Her husband, Wayne Hauser, was in his lab, and there was the connection with my PhD supervisor, Jim Watson. I arranged with both Alex and Susumu to spend time in Susumu's lab. There were some questions about how the cells I worked with in New Zealand recognized and killed tumor cells. I then bumped into David Baltimore whom I had met earlier in New Zealand. David had worked with Jim and Susumu at the Salk. When I mentioned my move, David said "Out of the frying pan into the fire". He did not smile. I hoped he was joking. He wasn't. Susumu seemed friendly enough, I thought, and maybe there was some history between him and David.

Susumu had previously shown that antibody diversity was generated by rearranging DNA at particular sites in the genome (right panel, Figure 3.3). This shuffling of segments allowed antibodies to recognize a huge variety of threats from the environment, even without ever encountering them beforehand. The discovery

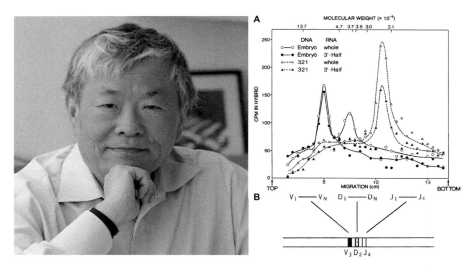

FIGURE 3.3 Susumu Tonegawa in 1986. The panel to the upper right (A) shows the very different size of the gene encoding the kappa light chain of an antibody in the embryo and in an antibody-producing cell, indicating rearrangement of the DNA. The size fractionation was performed using a gel made of potato starch and electrophoresed for three days. (Image from *Proc Natl Acad Sci USA* 73, pp. 3628–3632, 1976 (permission from Susumu)). The gel was sliced and the DNA hybridized with the radiolabeled RNA indicated. After digestion with RNase to digest the unbound RNA, the hybrids were precipitated and the radioactive counts from each slice were determined. The lower right panel (B) is from the Nobel Committee press release (https://www.nobelprize.org/prizes/medicine/1987/press-release/). Susumu notes that it inaccurately shows the rearranged segments in the RNA as being separated by a spacer. That is a mistake; the segments are contiguous.

by Steve Hedrick, one of the New Zealand Jim Watson's students from his time in the US, revealed that a similar process occurred in thymus-derived immune T cells. The question remained: how many other lymphocyte receptors were generated by rearranging genes? I was able to produce clones of the activated tumor-killer cells I had worked with in New Zealand. Each clone differed as to which tumor cells it preferentially killed, suggesting there was some specificity to this process due to the receptors involved in the interaction. Different strains of mice also killed a particular tumor better than others. Along with "Tak" Takagaki in Susumu's lab, we showed that the cells that I grew out did have a variable surface receptor that was not encoded by the known receptor genes. Intriguingly, it was a dimer that had a similar size to the known rearranging genes that produced T cell receptors. Then, the project fell apart. Susumu wanted me to just grow the cells and hand over the identification of the receptor to others. That would mean a buried authorship or just an acknowledgement that I supplied the cells. One explanation he gave was that it was unlikely that I had the manual dexterity to be a molecular biologist. Others in the lab already had the skills necessary. Susumu said that it was my use of chopsticks that betrayed me.

Another issue came up when I changed fellowships once the Fogarty Fellowship ended. Susumu said that there were differences with a new funding from the Cancer Research Institute but not to worry as he could help with those issues. It turned out my daughter needed ear surgery for recurrent otitis media. Much of the cost was not covered by the new medical insurance and we were barely paying our bills as it was. When I mentioned that to Susumu, his response was "If someone's grandmother dies, would you expect me to pay for a plane ticket so they could go to the funeral". As I returned to my bench, the phone rang and Alex asked "How're things going?". Timing is everything. It was time to leave and rejoin Alex's lab. No one in Susumu's lab was surprised that I left, as my exit was like many others and followed shortly after that of an Australian post-doc, Bruce Robertson. We certainly were not the last to make an unscheduled retreat. In fact, one section of the lab was jokingly called the departure lounge. It was the most distant part of the lab from Susumu's office at the end of a maze that connected separate parts of the lab.

There were certainly some cultural differences that led to problems. Susumu himself felt that he had not always been treated fairly in his early days at the Salk Institute. Jim Watson, my New Zealand supervisor, confirmed there were some clashes, including with David Baltimore. Susumu's visa expired and he went to Basel in Switzerland where he made his great discovery. He was awarded the Nobel Prize the year I left his lab. According to Ellie, his street-wise Boston secretary with a gravelly smoker's voice and much older than Susumu, he was pleased that he alone received the honor and did not share it with others who were also in the running.

Susumu's discovery was breathtaking. To show the gene rearrangement, Susumu had purified his own enzymes, isolated the RNA that encoded antibody proteins and separated DNA fragments hybridized to the RNA on potato starch gels that took three days to run (Figure 3.3) [32]. He then meticulously sliced the gel by hand to show that the radiolabeled DNA fragments migrated differently when isolated from B cells compared with liver cells. In the same year, Phil Leder, from Harvard, and Susumu shared the Lasker Prize. In his interview at the time, Phil Leder noted that scientific research was not easy. "If you can't stand failure and disappointment, you can't do this work", he said. "What you work for in this business is the grudging appreciation of the few colleagues who understand what you are doing".

I had not learned a lot of molecular biology with Susumu but he was a good teacher in other ways. One rather long session started with a discussion of one of my first results that was less than stellar. I was still relatively fresh off the boat, as they say. We looked at the data lane-by-lane, and the protocol step-by-step. We did that again, and then again, looking for what might be done differently. Susumu knew what the problem was, but was waiting for me to use the word that he wanted to hear. The word was "accident", as in "I made a mistake by accident", meaning it was not intentional. He then echoed the word "accident" very deliberately and asked, "Was it an accident?". I said I thought so. He then said, "Isn't an accident something you would not have anticipated?". That was the point. If you could anticipate that an event could happen, then it was not an accident. You just didn't plan well enough. For example, if you leave a glass on the edge of a counter and then you "accidentally" knock the glass and it smashes as it hits the floor, is that an "accident"? It took

Susumu an hour or more to make his point. This and other similar sessions led me to the conclusion that everything Susumu did was intentional – nothing was accidental. Of course, please don't use this definition of an accident with your friends or family. It will not work in the real world. Usually, sympathy is the best response. However, the approach does cause you to think about how you plan your experiments. It does make you hyper-critical of your own work. It does focus your attention on how to proceed rather than on an excuse for your failure. Running a bad experiment often takes much more time than planning one that will yield meaningful results. The process sets you free to take the next step forward. Only after you assess the situation in a very detailed way do you make your move.

And so, after the phone rang with Alex at the other end, I made my move, very deliberately.

4 From One Unknown to Another

Not much had changed in Alex's lab when I returned to take up the quest of finding a biological function for Z-DNA. There were two new post-docs, Nassim Usman and Loren Williams, who were quite a lot of fun. Loren was always threatening to become a bus driver; although it was just as aggravating as science, the pay was much better. Nassim was Canadian and an excellent RNA chemist. He was often invited for talks and always managed to fly to those occasions first class. He also liked Cuban cigars and was a coffee aficionado long before it was fashionable. Coffee hour became a must. Alex would often drop in when he arrived around 3 pm. Alex was fond of impromptu philosophical team talks. The topics would vary and highlight stories from the early days of molecular biology. One of his favorite reflections was on how scientists weren't really much interested in money and most managed to have quite comfortable, middle-class lives. In that era, the wealth of biotech millionaires was newsworthy. The listings from trade journals would mysteriously appear on the lab notice board. Alex was usually at the top of the list because of his association with one of the first biotech startups called Repligen. Later, because he was a proven winner with the investors, he was paid to join the board of many other startups. Truly, a case of the rich getting even richer. Alex would remove those postings describing his wealth within a very short time of their appearance. For those of us struggling to pay Boston rents and raise children, it was amusing to see how scrupulously he took those notices down so that we would not be distracted.

Alex was a post-doctoral fellow at Caltech with Linus Pauling for a number of years. There, he met quite a famous cadre of scientists including the American James Watson, Max Delbrück, Carleton Gajdusek, and Richard Feynman, along with many others of distinction, including Irwin Oppenheim, Verner Schomaker, Jerome Vinograd, Hardin McConnell, and Benoit Mandelbrot. Apparently, Alex was well known at the time for his parties and vineyard tours but was a teetotaler by the time I knew him.

His project with Linus was to record the X-ray patterns of DNA fibers, much as Wilkins, Gosling, and Franklin were working on in London. It seems that his apparatus was of low intensity and not ideal for producing high-quality images. Alex always said that if Linus had seen Franklin's data, he would have been the first to propose the correct model for DNA, just as he had been first to deduce how amino acids fold to produce a globular protein structure. However, the Pauling-Rich DNA model was not to be (Figure 4.1) [33]. Apparently, there were hints of the double helix in those images that Alex was able to capture. Alex was no Ray Gosling. Nor akin to Rosalind Franklin. Maybe...if... – I wouldn't even consider it a race.

DOI: 10.1201/9781003463535-5

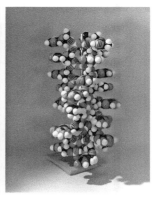

FIGURE 4.1   Linus Pauling and Alex Rich and their beret club. The Pauling-Rich model of DNA was never to be. Shown is the triple helix DNA model proposed by Pauling and Corey with the bases pointing outwards (image from https://paulingblog.wordpress.com/2017/01/28/the-triple-helix/).

There never was a Rich and Pauling publication, then or in the following years. Alex liked to tell the story that he once considered himself a failure and that he was no good at science. Alex would often tell the story about Pauling saying that "He didn't know what Alex did when he was in his [Pauling's] lab, but he must have learned a lot". Paul Schimmel, another MIT professor and also on the same biotech boards as Alex, related that same story in his *Nature* journal obituary for Alex [34].

As they say in academic circles, Alex "managed his career" well. He joined the Public Service at the NIH during the Korean War and did a sabbatical in Cambridge, England while his lab was being built. He was a founding member of the RNA tie club that included Watson. Crick, Brenner, Orgel, Gamow, and others (Figure 4.2). Six of the twenty members won Nobel prizes. The club was a celebration of the cracking of the genetic code. Each member had a tie with a cartoon representing one of the amino acids. Alex was quite proud to belong. He was arginine and his title was "Lord Privy Seal of the British Cabinet". Not bad for a kid who grew up in a working-class family in Springfield, Massachusetts.

Alex moved to MIT in 1958. One of his focuses was on the structure of transfer RNA (tRNA). The tRNA molecule is the small RNA adaptor that maps the 3-letter genetic code (called a codon) to a particular amino acid (Figure 1.7), enabling the production of a protein from a messenger RNA. How did it connect a codon in messenger RNA to the correct amino acid? His lab was eventually able to obtain crystals that led to a structure published in 1973 in the journal *Science*. The work was done by Sung-Hou Kim, who left to go to Duke before the high-resolution structure was finalized.

The publication of the MIT/Duke paper on the tRNA structure led to a dispute that was featured in the September 19, 1974 issue of the *New Scientist* magazine: "Transfer RNA researchers argue about borrowed data". The issue was raised by the crystallography group led by Aaron Klug in Cambridge, England, whose article

FIGURE 4.2   Alex Rich's favorite picture with some members of the RNA tie club. From left to right are Francis Crick, Alex Rich, Leslie Orgel, and Jim Watson. Francis, Alex, and Jim are wearing their RNA ties (from https://paulingblog.wordpress.com/tag/rna-tie-club/).

appeared two weeks later in *Nature*. Francis Crick was cast into the role of mediator. Both the Rich and Klug lab presented their structures at a scientific meeting prior to publication. The English group claimed that the structure Alex presented at the meeting were wrong and that he used their results to correct his model before beating them into press. Sung-Hou's structure actually differed from the one Alex presented at the meeting. Alex did not know much about that because of "a breakdown of communication between the MIT and Duke groups in 1973–1974" (from Sung-Hou's Wikipedia page). It seems that the MIT and Duke groups rapidly reestablished contact once they realized that the English group would beat them to press [35]. As Andy Wang recalls, "There were intense negotiations between Alex and Sung-Hou in how to publish the paper. In the end, Sung-Hou was put as first author and Alex was the corresponding author". (email to AH, April 11, 2021). They were first to publish [36].

The Klug group was not pleased with coming in second, given that Alex's published structure was different from the one he described at the conference, and now was undeniably very similar to their model. Given the circumstances, the Marquess of Queensberry Rules were suspended. The scientific slugfest was savage and there was never a resolution. The *New Scientist* article created fractures that were never healed. Clearly the nature of "borrowed data" was considered by the Cambridge group in England to be different from the "borrowed data" that was key to modeling the DNA structure by their Cambridge colleagues Watson and Crick. That dynamic duo did not even try to derive any data experimentally but instead had the use of unpublished data from the London group. There was a protocol for such situations that those in the club knew to follow. Apparently, what Alex did was judged so egregious that he was pretty much on the outside from that point on. All Sidney Brenner would say to me about Alex was stated in a rather muted voice that only an insider would use to describe the flaws of an erstwhile colleague: "Alex did things he shouldn't have".

Contrary to the oft-quoted advice, the Club rules were quite simple. Follow protocol, and ask for permission rather than expecting forgiveness. For the English establishment, a wink and a nudge, say no more, is often all the complicity you need. Furthermore, if you don't feel the need to seek forgiveness and insist that you are innocent, then they will likely respond with the words of Willian Shakespeare: my Lord Privy "… doth protest too much, methinks". Other fallen angels, including Bob Gallo, Carlton Gadjusek, and Carlo Croce, could always count on Alex to rally to their defense. The publication of the tRNA structure was pretty much a dead tie as scientific races go. The spat detracted from the stunning achievements of both groups. It was at that time the latest RNA structure ever solved, by a long way!

In these stories, it is always a little hard to sort fact from fiction. Even Sung-Hou and Alex had different recollections, with the letters between Alex and Francis Crick providing additional context. Those letters are to be found in the Francis Crick archive at the NIH. They were made public in 2003 by the Wellcome Trust to mark the 50th anniversary of the Watson-Crick DNA model. The exchanges, which would not have persisted in this day of rapid email exchanges, allow the reader to see how the events were portrayed differently by the warring parties. Prior to seeing these carefully crafted documents, I was never sure that the history Alex told would be the same one that others recalled.

A different example relates to the discovery of Z-DNA. The electron density maps obtained from the crystals were not consistent with B-DNA. In that era, before the age of computer graphics, the maps were printed on transparent plastic sheets and layered between plexiglass squares. This process was repeated to create a stack of the electron densities, mirroring the distribution of atoms in the crystal. Then, the challenge was to fit the DNA molecules into the electron density map so that there was a good fit. Making derivatives of the DNA with heavy metals increased the intensity at certain places in the electron map, providing helpful hints as to the location of the residues bound by the metal. Of course, everyone was expecting a right-handed helix to be the correct answer. It was not to be. So, who first thought to restack the plexiglass electron density plates to deduce that the DNA helix was left-handed? Was it Andy Wang or Alex Rich or someone else like Gary Quigley, who wrote the Fortran programs to calculate the electron density maps in the first place? My bet is on Andy, who replied to me when I asked him that question: "Well, Alex was told by me that the stacking direction of glasses … should be reversed in order to get the correct structure solution … in the corridor of Building 16, 7th floor…in a late night of 1979 summer. I did not turn the glasses upside-down; I just reversed the stacking direction. Alex initially did not like my suggestion of naming it Z-DNA, because Z was in last place of the 26 [letters of the] alphabet. But neither R-DNA (for Rich), L-DNA (for left) nor W-DNA (for Wang) seemed proper, so he reluctantly accepted it, especially after someone from Germany said to him "Das ist ein good idea"! But Alex is Alex, so I left Alex [to] enjoyed (sic) most of the glories."(email to AH, April 11, 2021) (Figure 4.3)

FIGURE 4.3   Andy Wang and a model of Z-DNA (adapted from https://web.nstc.gov.tw/SciencePrize/2021/4910936633.html).

Over time, Alex ended up on many high-profile committees and was elected to a number of academies in different countries, including the Pontifical Academy of Science. Also, he helped a few Russian scientists make the transition to and become established in the United States, much as the earlier generations of his family were helped to leave Russia in the same fashion. I am not sure of the full list but Alex Varshavsky and Maxim D. Frank-Kamenetskii come to mind. Alex also preferred to keep people around. Sung-Hou Kim was with Alex for six years. Both Andy Wang and Gary Quigley, who solved the Z-DNA crystal structure, were around for over 14 years. The long-term positions were not how other labs staffed themselves. The usual post-doctoral associate trains for two to three years, usually exiting when their fellowship funding runs out. Everyone had their own reasons for staying. Of course, timing is everything. I was made aware of a position for me back in New Zealand. As it was, I had not been given a heads-up on this position and had just sold my house in New Zealand so I could buy one to fix up in the Boston area; I felt like I may have been their second (or third) choice. I was given an exit back to New Zealand but had still not finished the job of finding a Z-DNA binding protein. It was not a failure that I wanted to live with.

Both Alex's connections and the productivity of the crystallography group led by Andy and Gary helped Alex remain funded. His grant from the Office for Naval Research (ONR) ran for many years. Ostensibly, the money was to crystallize a light-absorbing pigment from bacteria that might have an application to stealth technology on ships. There never seemed to be any active work in the lab on this project that I was aware of, although Gobind Khorana's lab, also at MIT, worked on this protein. The structure was finally solved in 1998, by someone else. Not a great day for Alex. Anyway, Alex had a great relationship with his ONR program officer, who was sure to phone Alex if the application for renewal of funding had not been received in time. I was grateful as my salary came from that grant. The ONR was also helpful when it came to visa renewal. Being part of the Defense Department, they were not subject to the same restrictions in extending visas as my original sponsor, the National Institutes of Health (NIH). Not all votes in Washington are created equal, and, with the ONR thumbs-up, I was eventually able to progress to the more permanent status of Resident Alien.

Once back in Alex's lab, I started my search for Z-DNA-binding proteins. The first order of business was to clean up ... a lot. One thing Nassim and I found in the process was a lump of sodium metal just sitting there, exposed to air. It just sat there at the back of a fume hood stacked full of many explosive organic chemicals in gallon flasks, from some of which the labels had fallen off. It was fortunate that we found the sodium, given the danger it posed. The sodium metal had been there such a time that it was bare-naked, no longer submersed in oil or kerosene or whatever was originally used to prevent its exposure to water. Had water touched the sodium, then the heat generated may have been sufficient to cause a nasty outcome. Nassim took care of the disposal. After all, he was the chemist in the group. He dumped the sodium in a sink full of water, causing a large, but contained eruption that emptied the lab. Nassim was pleased with his handiwork. Another post-doc barely escaped serious injury as he took a door off its hinges as he made his hasty exit. I am not exactly sure why the sodium was there, but the rumor was that one of the former Rich lab members was into manufacturing some illicit psychotropic drugs to finance his lifestyle. Who knows?

When it came to Z-DNA-binding proteins, no one knew what to look for, of course, assuming that Z-DNA-binding proteins existed. At that stage, there was no evidence for the existence of anything that specifically bound Z-DNA apart from antibodies Eileen Lafer had made by immunization with the brominated polymer. If Z-DNA-binding proteins did exist in the cell, would the proteins bind to specific Z-DNA sequences or just to the Z-DNA conformation? Specificity of binding was the possibility favored by Alex, since the base-specific residues were exposed on the convex surface of Z-DNA and protruded out from the helix (Figure 2.4). They are not buried in a groove like they are in B-DNA and A-RNA. The surface of Z-DNA is information-rich, compared with these other structures.

Another question was how do you find the Z-DNA-binding proteins? Initially, Achim Möller's bromine-modified poly-d(CG) was used to purify proteins. The polymer was very long with many potential binding sites for proteins so it likely would pick up low-affinity interactions involving protein complexes, with patches of positive charges that would bind to the negatively charged Z-DNA backbone. The length of the polymer and the strength of the multiple G:C bonds present allowed it to form hairpins and structures where multiple strands hybridized to each other. There was no way to know what the proteins were binding to with this substrate. Was it the chemical modification, the regions of single-stranded DNA that were present, or some other DNA structure, or were they recognizing Z-DNA?

The modified polymer was also used to assay proteins eluted from the column made from the bromine-modified poly-d(CG). Proteins that bound the polymer would trap the modified DNA on a glass filter. The amount of binding could be determined by using radioactively labeled DNA. Although many proteins bound, none were ever shown to be Z-DNA-specific as there was no way of telling. It became a game of semantics. If a protein bound to brominated d(CG) column polymer, then it was, by definition, a Z-DNA binder, even if there was no subsequent characterization in terms of structure or function. The protein zuotein (named because zuo equates to left in Chinese) is one example. Subsequent work demonstrated a role for zuotein in the correct folding of proteins soon after they were translated by the ribosome. One interesting spin-off from zuotein was the realization by Shuguang Zhang that zuotein contained regions of repeating positive and negative charges capable of assembling into fibers and sheets [37]. He found this out when he chemically synthesized the sequences as peptides. He was initially interested in seeing whether the peptides bound to Z-DNA but switched his research direction when he discovered their unexpected properties that allowed him to form peptide fibers and float them on water to form membranes. Because these peptides are biodegradable, they have found clinical application as topical hemostatic agents for controlling bleeding during surgery. That application was a commercial success and made Shuguang a wealthy man. The success was far from the field of Z-DNA, but an illustration of how science produces unexpected outcomes. An incredible story given that Shuguang survived the agrarian reforms implemented by Mao Zedong during the Cultural Revolution.

It was a dismal time for anyone involved in Z-DNA research. There were many publications from a multitude of groups showing that any sequences capable of forming Z-DNA inside the cell were either mutagenic or increased the risk of cancer, suggesting that Z-DNA was a problem, not something nature would embrace. There was even one paper claiming that anything that bound Z-DNA was really something whose real function was to bind negatively charged phospholipids that somehow might look like Z-DNA. Then, there were the conferences with the poster sessions where people would take a glance at your presentation out of the corner of their eyes, put their shoulders back, noses up, and resolutely walk by. Everyone seemed convinced that Z-DNA was a lost cause.

Many of the ready-made projects Alex had promised his newest recruits as ongoing in his lab didn't exist or never yielded the results he described when recruiting them. Many of the arrivals just left as I originally did. Others, like Loren, changed their focus to crystallography where the chances of a paper were higher. Even that was a strategy marred by the Rich lab culture. With the crystallographers, Alex would set them off on competing projects where each would use related DNAs or compounds to make crystals. Alex would be the middle man passing on results from one to the other as suggestions that had just come to his mind, with the inevitable result that people felt played off against one another. These were smart people as shown by their credentials. Andy Wang, while still with Alex, would often mediate and would help as Andy really liked to solve structures. Once Andy was gone, there was no buffer. The crystallographers often ended up not liking each other.

Nassim did a lot of work for other groups. He helped Paul Schimmel synthesize mini tRNA helices that at one time made the press as a "second genetic code", an idea that faded quickly. Nassim also aided Jennifer Doudna when she was with Jack Szostak at Massachusetts General Hospital. Jennifer went on to win the Nobel

Prize for CRISPR. Another student to whom Nassim taught the art of RNA synthesis formed a company that eventually was sold for a lot of money. Nassim was also the chief scientific officer for Ribozyme Pharmaceuticals, a company founded by Tom Cech, another Nobel Prize winner, but left and sold his stock after coming out on the wrong end of some internal politics. A few months later, the company changed its named to SIRNA and sold itself to Merck & Co. for a billion dollars. Nassim then spent some time as an entrepreneur-in-residence before taking on a leading role in a biotech start-up that focused on the protein therapeutic space.

For a while, I was the only one working on assaying Z-DNA binding proteins. Ky Lowenhaupt, who ended up spending over 20 years in Alex's lab, was helping with protein preparations, with equipment built up over the previously unsuccessful campaigns. Ky helped everyone and liked to be at the center of things. The initial attempts to purify proteins were based on the long polymers stabilized in the Z-DNA conformation. Bacteria, yeast, and other easy-to-grow organisms were studied first. There were proteins that bound to the column and were enriched in the assay. The results led to many publications with no follow-up. A series of one-shot wonders, as it is known in the trade, a sign that the underlying results are not robust. Ky would, of course, know what the inside story of what happened and would share her thoughts freely on the subject. Others would turn to Ky, especially any new members of the lab, for help and advice.

As my first steps, I tried the polymer approach to purify and assay for Z-DNA-binding proteins. Not unexpectedly, the approach failed. It was time to try a different strategy. When stuck like this, I always think back to Julius Axelrod's book that I read at medical school. His Nobel Prize was for elucidating the role played by neurotransmitters in regulating biological clocks. His emphasis was on developing a method best suited to solving the problem at hand. Following that advice, I developed a method to find Z-DNA-binding proteins using a very short probe that overcame many of these problems associated with the long polymers. The approach was inspired by the assay developed at MIT by Francois Strauss and Alexander Varshavsky that had been refined by Harinder Singh in Phil Sharp's lab. In particular, I could visualize binding interactions by separating the protein–DNA complexes in a gel. The approach was called gel electrophoresis and separated any complexes formed between DNA and protein by differences in size and charge. The position of the complexes in the gel could be visualized by a radioactive label in the DNA using X-ray film to image the radiation (Figure 4.4). The important part of the assay was the ability to compete with different unlabeled DNAs to find which kind of DNA was bound by the proteins in the complex. That way, I could check the specificity of binding with unlabeled B-DNA and Z-DNA polymers. I could also use plasmids from bacteria that, if made in the right way, could stabilize inserts in the Z-DNA conformation without chemical modifications to the DNA. I validated the approach by using anti-Z-DNA antiserum developed by Eileen Lafer working with David Stollar. The bands higher in the gel were the DNA bound by the antibody. Only the Z-DNA-containing plasmid would inhibit binding, showing that the assay was suitable for detecting Z-DNA binding proteins [38]. Most importantly, the method worked!

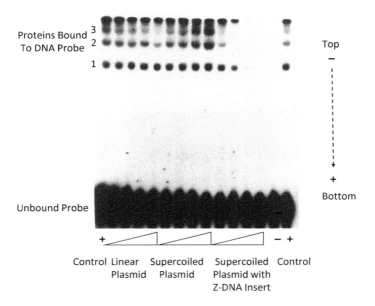

FIGURE 4.4   Assay for Z-DNA-binding proteins using radioactively labeled DNA and competition with the unlabeled DNAs named below the image. Here, the protein used was a goat anti-Z-DNA antiserum with different amounts added to each lane; the levels of protein increase from left to right. The image is from an X-ray film exposed to the gel after electrophoresis had been performed to separate free from bound radioactively labeled probe. The gel is run from top to bottom (negative to positive). There are multiple binding sites on the probe as indicated on the left but only an unlabeled Z-DNA containing plasmid competed for binding(see *Nucleic Acids Res*, 21, pp. 2669–72, 1993 for details).

It took a long time to convince myself that any Z-DNA-binding protein existed in a normal cell. First of all, we didn't know where to look for such a protein. Was there a particular organism we should use or a particular tissue? If the Z-DNA protein were sequence-specific, would it still bind to the probe? I tried many different proteins previously purified by Bruce Albert's lab in San Francisco and by Rick Fishel at Fort Detrick, but found nothing that looked like it was specific for Z-DNA. Maybe I had made the criteria too stringent? Or maybe I was wasting my time? Bottom line, there was nothing publishable. Why keep trying? At this stage, looking for a fast exit was the obvious choice. So, when is the best time to stop panning for gold? There are a lot of answers to that question. The answer I was getting from many was that it was time to pack up the mule and move on. Yet, the assay was robust. Was I prospecting for riches in the wrong places?

The fact that Alex's lab was not a biology lab was also a problem. Which cells could we use to isolate a Z-DNA-binding protein from? We were not set up to do large-scale cell culture. Nor were there any abattoirs in Massachusetts from which to obtain animal tissue. At that time, Jeff Spitzner joined the lab. He suggested we try chicken blood as the red blood cells were nucleated and he had used them to study topoisomerases. He arranged to collect blood at no cost from a chicken farm in Rhode Island, some 60 miles south of Boston. The initial results looked promising. There was a band we could see in the assay that would disappear when unlabeled Z-DNA was included in the incubation mix, but would not be affected if we used unlabeled B-DNA instead. By using ultraviolet light to cross-link the protein to the radioactive probe, then using nuclease to trim the DNA fragment to as small a length as possible, I identified a protein of about 41,000 molecular weight that bound to labeled DNA. I named the protein Zα [39]. Why? In those days we used Letraset sheets that allowed you to transfer with a pencil a typeface letter onto a figure to give a perfect label. Zα had two letters that were always available in the Letraset collection used communally by the Biology Department. The "Z" was for Z-DNA binding and the "α" implied that there was a "β" and a "γ".

Tiring of the drive to Rhode Island, and with us wondering what would happen if we were pulled over by the police covered in large splotches of chicken blood, Jeff then checked out the local Boston Chinatown, where live poultry was also processed. It seemed that the only tissue that we could obtain in sufficient quantity was lungs because this part of the chicken was not eaten. By this stage, I had found that the Z-DNA-binding activity pelleted in a high-speed centrifuge along with the ribosomes that made protein. This was promising, as we were looking for something that was likely part of a nucleic acid-protein complex. The activity could be eluted off the pellet in a high-salt wash. The protein concentrate was dialyzed, using a special membrane that allowed the removal of salt but not protein, then handed to Ky, who performed further purification. Ky passed the crude mixture of proteins over columns that separated proteins by size and by their charge. I followed the activity by testing each fraction in the band-shift assay. We were lucky that the activity seemed to be due to a single protein rather than a complex. The final step was to use an affinity column made of brominated Z-DNA, hopefully, to finally purify an authentic Z-DNA-binding protein.

It was not fun to make that brominated Z-DNA. Sounds simple. Take DNA, add water saturated with bromine gas, let it stand, stop the reaction by bubbling out the gas, and then check for the presence of the characteristic spectroscopic signature of Z-DNA showing that 30% of the guanines are modified. The process was crude and reminded me of the use of chlorine gas in World War I. Of course, the reaction was done in a fume hood, but those plumes of brown gas look as noxious as you can imagine. Repeat until the Z-DNA signature is optimal. If done correctly, you had the Z-DNA you needed to separate the protein of interest from those that had no affinity for Z-DNA.

To follow the purification, fractions were run out on a protein gel to see what we had. I tracked those that bound the Z-DNA probe to see whether a particular protein band copurified with the binding activity. There was a band of 150,000 molecular weight (p150) that correlated with Z-DNA-binding activity. I directly confirmed that a protein in this region of the gel was the one binding to Z-DNA by a technique called a Southwestern blot. Here the protein was first separated by gel electrophoresis. The gel was then laid on a nylon membrane and the protein transferred by blotting from the gel to the membrane. In this step, a voltage gradient was applied perpendicular to the original direction of electrophoresis. The membrane, now with the protein bound to it was exposed to radiolabeled probe to see which protein bound it. There was no reason to believe that this method would work. To separate the proteins by molecular weight, it was necessary to boil the protein in a detergent that had negative charges at one end. The idea was that the detergent molecules bound uniformly along the protein backbone to linearize them, allowing them to be separated by their length. The negative charge ensured the proteins would move in the same direction in the gel. In short, the whole process destroyed the folding of the protein and any activity it might perform. But the assay worked! The p150 did bind the protein and the binding was competed for by unlabeled Z-DNA. The results all looked very promising [40].

The purified protein was sent for sequencing to determine the order of amino acids (Figure 4.5). The technique at that time was not very sensitive and a lot of protein was required. However, we obtained two unique peptide sequences. They matched with a high degree of certainly to a human protein that had recently been cloned. That sequence had just been released to the database of protein sequences maintained by the NIH. Good timing for us! We did not have to clone the chicken gene but could use the related human gene. I contacted Kazuko Nishikura from the Wistar Institute in Philadelphia, whose lab had just isolated and cloned the human sequence [41]. She sent us the protein expressed in insect cells containing segments of her cloned DNA! We were in business.

The protein produced by the DNA fragment encoding the first 196 amino acids of the sequence gave a signal in the Z-DNA assay whereas other fragments did not. The band was competed with the unlabeled Z-DNA! We had the right protein. With the DNA clone in hand, I rapidly mapped which of the 196 amino acids resulted in Z-DNA binding by expressing different DNA fragments in bacteria. I had never cloned anything before, especially not in Susumu's lab. We had the right protein and the activity was so robust that even a novice like me could map the binding domain! I did not need to use chopsticks (see chapter 3)!

But what did the protein do? The human protein was rather mysterious. It was one that had been isolated because of what it did to double-stranded RNA (dsRNA).

Table 1. Comparison of dsRAD sequences

Peptide 1		LQAPYQINHPEVGRVSVYD
		I+ I+ +III IIII I+II
H-dsRAD	1095	LRHPFIVNHPKVGRVSIYD1113
R-dsRAD	1041	--Y------------V--1059
Peptide 2		K()()()RIFPAVTA
		I I+IIII+I
H-dsRAD	763	KVGGRWFPAVCA774
R-dsRAD	709	------------720

FIGURE 4.5 The two peptide sequences we obtained from the chicken protein was a match to human and rat ADAR1 (then known as dsRAD for double-stranded RNA adenosine deaminase). The vertical lines represent a match and the + a conservative amino acid change. Herbert, A., Lowenhaupt, K., Spitzner, J., Rich, A. *Proc Natl Acad Sci USA*, 92, pp. 7550–7554, 1995) (Copyright (1995) National Academy of Sciences, USA).

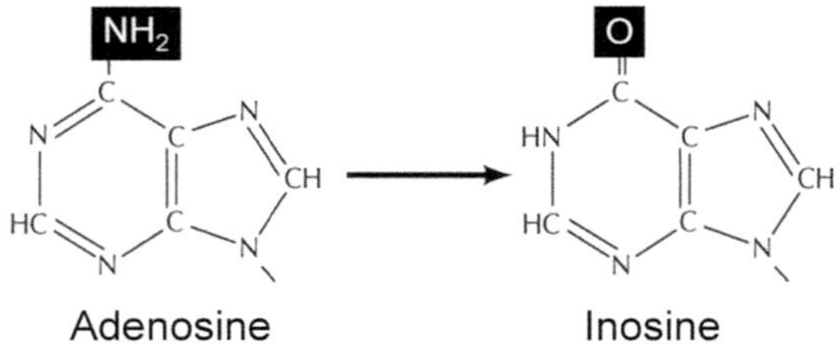

Adenosine   Inosine

FIGURE 4.6   The adenosine-to-inosine conversion catalyzed by the RNA editing by ADAR
(Adenosine Deaminase Acting on RNA) that replaces an amino group with a keto group,
resulting in the deamination of adenosine. Bass BL, Weintraub H. An unwinding activity
that covalently modifies its double-stranded RNA substrate. *Cell*. 1988;55(6):1089–98. doi:
10.1016/0092-8674(88)90253-x. PubMed PMID: 3203381.

It enzymatically changed one specific RNA base into another (Figure 4.6) [42]. The
protein did that by replacing an amino group (NH_2) on adenosine with a keto (O=)
oxygen to give inosine.

The amino group of adenine was one of those involved in Watson-Crick base-
pairing (Figure 1.6). So, replacing the amino group with oxygen to make inosine
meant that the base no longer paired with thymine. Instead, the modified base preferred to pair with cytosine. The adenine is thus replaced by a base equivalent to
guanine. The modification changed the genetic code of the RNA so that a different
amino acid would be inserted into a protein rather than the one specified by the DNA
(Figure 1.7 to see how). The process is called RNA editing. The replacement of one
amino acid by another is called protein recoding.

Okay, what does Z-DNA have to do with RNA editing? It was necessary to show that
the chicken p150 protein we purified did have the same A to I enzymatic function. The
first step was to rummage through the lab to find amongst the piles of discarded equipment just what I needed to run the deamination assay. The method was old school. The
required pieces were found without too much trouble. I needed some thin-layer chro-
matography plates and I found some of unknown vintage. These plates were designed
to separate the different bases from one another. They were suitable for resolving the
radioactively labeled adenosine from the inosine produced by RNA editing. Then, to
run the assay all that was necessary was to incubate the purified Z-DNA binding protein with a dsRNA that contained many radioactively labeled adenines. After that, the
single nucleotides (base-sugar-phosphate) could be obtained from the dsRNA by using
an enzyme called RNase to cleave the backbone holding everything together. Under
the conditions used for the thin-layer separation of the nucleotides, the adenosine and
inosine spots moved to different locations in the plate. It was the first time that I had ever
performed this experiment. You do what you have to do to answer the question at hand.
The protein certainly catalyzed the editing reaction! Another rookie success!

The next thing was to test whether the protein really bound in a Z-DNA-specific
manner. I wanted to use a technique unrelated to the assays I had used for the purification. It would be best if I could use the recombinant protein that I had made in
bacteria. By employing this material, I would rule out the possibility that other factors in the chicken lungs were contributing to the Z-DNA binding I had observed.

For this work, I would need milligrams of pure protein. Again, luck was with me as the Z-DNA-binding domain was extremely stable and expressed well in bacteria. I could make the large quantities of the protein I required [43]!

My aim was to use the original assay Pohl and Jovin had employed to find the first evidence for the existence of left-handed DNA. This approach is based on circular dichroism. The method can distinguish between right-handed B-DNA and left-handed Z-DNA by testing whether right- or left-polarized light is absorbed by the d(CG)$_n$ polymer. In this case, I used the 5-methylated (me^{5}dC-dG) polymer that exists in the B-DNA conformation in 50 mM NaCl and is prone to flip to Z-DNA when low concentrations of certain metal ions are present. I had no idea as to whether the experiment would work. I could imagine many reasons why it would fail. The procedure involved adding a small amount of the Zα protein to the polymer in an optically clear glass cuvette and then taking a measurement. The Zα protein preparation I used was pure, of course, and free of any metal ions. I did not want to fool myself by using poor-quality reagents.

The Jasco machine for making the measurements was set to scan wavelengths between 230 nm and 320 nm. The process takes about one minute to complete per wavelength. The results are displayed one slice at a time on a cathode ray tube screen. As each scan is completed, a new point is added to the line on the display screen (Figure 4.7). The change I was looking for only started to appear at 290 nm, but I had to wait until the lower wavelengths were measured and those results were processed. It was like watching a frame-by-frame replay of a foot race – everything served up in slow motion with the finishing line hidden until the very end.

The first addition of Zα to the polymer produced no discernible effect. Then, on the second addition of protein, it looked like the spectrum shifted. Maybe the protein

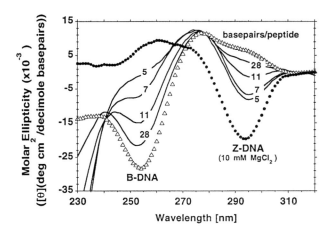

FIGURE 4.7 The flip from B-DNA to Z-DNA induced by the Zα domain. Herbert A, Alfken J, Kim YG, Mian IS, Nishikura K, Rich A. A Z-DNA binding domain present in the human editing enzyme, double-stranded RNA adenosine deaminase. *Proc Natl Acad Sci USA.* 1997;94(16):8421–6. Epub 1997/08/05. PubMed PMID: 9237992; PubMed Central PMCID: 22942.

itself was causing the change? I had not yet scanned the protein to see whether it was optically active so I didn't know. However, I was not looking for just any change. I was looking for the change Pohl and Jovin found, where a strong negative peak around the 296 nm wavelength would occur. That change is specific for Z-DNA. With the third addition, the shift was clear. I waited and then repeated the scan, then I repeated it again and again. The 296 nm deflection increased with each subsequent measurement (Figure 4.7). Eventually, the shift maxed out. Zα had flipped the polymer to Z-DNA!

I was more relieved than excited. My reaction was pretty clinical as, of course, this was what I would expect if the protein was binding to Z-DNA. Didn't I know that already? So, what was the surprise? It was just like you know when you cross the finish line that the race is finished. But you still need to wait until the judges finally call the result. It took a while to check the controls with protein alone and to do those using metal ions see how the protein-induced flip compared. Then, I had the confirmation I needed.

However, it was not yet time to celebrate. There were still a lot of unanswered questions. Just the same as when a climber gets to the summit of a mountain. The adventurer takes a photo to show that they made it to the top, then worries about how to make it safely down. If they don't, to paraphrase the words of Sir Edmund Hillary, it doesn't count. Take that, George Mallory and Andrew Irvine. By the way, whatever happened to their camera with that shot they took on the summit of Everest? Maybe their last conversation was along these lines: "Andy, what do you mean that you *accidentally* forgot to pack the camera?", then, after a brief pause, "Andrew, how can you call that an *accident*?" , just as Susumu might have asked had he been there (see chapter 3). I still had work to do.

One question was whether the RNA-editing enzyme was the only protein with a Zα domain. Perhaps there were more clues to the biology there. The way to test for this was to look for other proteins that had a similar sequence. I identified what was later renamed as DLM-1, then ZBP1, then DAI, then ZBP1 again (the mouse expressed sequence tag (mEST) in Figure 4.8) and also a protein from the vaccinia virus called E3. It also flipped the methylated polymer to the Z-DNA conformation after an overnight incubation. (Figure 4.9, unpublished data). I also identified a related sequence in the ADAR1 RNA-editing enzyme I called Zβ. I obtained the mEST clone for ZBP1 and showed that this could bind Z-DNA, but that the Zβ domain did not. A construct was made with the Zβ domain but did not produce a band shift in the assay. I then fused the Zβ domain to the Zα. Perhaps Zβ bound Z-DNA less tightly than Zα did but would bind more tightly if Zα held the probe in the Z-DNA conformation. In addition to the Zα + Zβ construct, I also made a Zα + Zα control. When compared with Zα + Zα, only every second band had Zα + Zβ present. The result suggested that Zβ could promote dimers of Zα (the reason why every second band was missing), but did not actually bind to Z-DNA. The reason for that result would become apparent once we had the crystal structure.

With the help of Saira Mian, whom I had met at a Gordon Research conference, we looked at how all the Zα-related sequences aligned. From the properties of the

A.

```
 1_hza___76    GVDCLSSHFQELSIYQD.QEQRILKFLEEL GEGK.ATTAHDLSGKL GT PKK.EINRVLYSLAKKGK LQKEAGTPPLWKI
 2_rza___76    GAEGLCSHFQELSISQN.PEQKVLNRLEEL GEGK.ATTAYALAREL RT PKK.DINRILYSLERKGK LHRGVGKPPLWSL
 3-bza___76    GVDRLSSHFQGLTISQD.QEQRTLELLDEL GDGK.ATTARDLARKL QA PKK.DINRVLYSLAEKGK LHQEAGSPPLWRA
 4_xza1__76    YIHSLSQAFGSLTVSHDILENNLLTFFKEI G.TK.TFTAKALAWQF KV EKK.RINHFLYTFETKGL LCRYPGTPPLWRV
 5_xza2__77    YIHSLSQAFGSLSVSRDPLENILLTFFRGQ GDTQ.TFTAKALAWQF KV KKK.HINYFLYKFGTKGL LCKNSGTPPLWKI
 6_hzb___72    TSALEDPLEFLDMAE-.IKEKICDYLFNV S--D.-SSALNLAKNI GL TKARDINAVLIDMERQGD VYRQGTTPPIWHL
 7_rzb___72    ASDLEGPSELLDMAE-.IKEKICDYLFNV S--K.-SSALNLAKNI GL AKARDVNAVLIDLERQGD VYREGATPPIWYL
 8_bzb___72    PCGLEEPPEPLDMAE-.IKEKICDHLFNV SS--.-SSALNLAKNI GL TKARDVNAVLIDLERQGD VYRQGTTPPIWYL
 9_xzb1__72    SEDTLITCSPEDMAG-.NKEKVCEFLYNS PP--.-STTLIIRKNV GI SKLPELNQILNTLEKQGE ACKASTNPVKWTL
10_xzb2__72    SEDTSVTSSPEDMAT-.NSAKVCEFLYNS PP--.-STPFIIRKNV GI SKMPELTQILNTLEKQGE ACKASTNPVKWTL
11_e31___68                  MSKIYIDERSNAEIVCEAIKTI GIEG.-ATAAQLTRQL NM EKR.EVNKALYDLQRSAM VYSSDDIPPRWFM
12_var___68                  MSKIYIDERSDAEIVCEAIKNI GLEG.-VTAVQLTRQL NM EKR.EVNKALYDLQRSAM VYSSDDIPPRWFM
13_mEST__67                  MAEAPVDLSTGDNLEQKILQVL SDDGGPVKIGQLVKKC QV PKK.TLNQVLYRLKKEDR VSSPE--PATWSI
```

B.
```
Multilevel                                          [STALxLAKNL GV PKK] [INRVLYDLERKG]      [GTPPLWxL]
consensus                                           [T    A   PQI L ] [V  I I    RQ ]      [T        ]
sequence                                            [          K    ]                       
(MEME)                                                        MOTIF I         MOTIF II      MOTIF III
```

C.
```
DSC_SEC                         CCC.HHHHHHHHHHHHH CCCC.HHHHHHHHHHH CC HHH.HHHHHHHHHHHCCH HHCCCCCCCCCCCCC
PHD_SEC                         ....HHHHHHHHHHH. .....HHHHHHHHHHH .. HHH.HHHHHHHHHHH... ...E........EE..
                                        HELIX A          HELIX B            HELIX C
```

D.
BLOCK (LAMA)	alignment length	score	Z-score score	expected value		
HTH_ICLR	(25)	(39)	(8.1)	(0.0e+00)	LTELAQKA GL PKS.TVHRLLQTmqqcgf	v
HTH_GNTR	(36)	(29)	(7.9)	(0.0e+00)	sERELAEEF GV SRT.TIREALRqleaegl	verkqgsgtfv
HTH_CRP	(38)	(28)	(7.3)	(7.5e-03)	tRQDIADYL GL TRE.TVSRLLGrlqeegl	isihgkrivi
HTH_DEOR	(34)	(30)	(6.9)	(2.5e-02)	SVEELAALF GV SEM.TVRRDLneleeqgl	lmrthgga
HTH_LACI	(32)	(21)	(6.7)	(3.5e-02)	TLKDVARLA GV SKS.TVSRVLnnnskvse	etrerv

ZA_BLOCK   TAxQLAKNL GV PKK.EINQVLYDLERQGK VYKSSGTPPLWSL

FIGURE 4.8   Our alignment of Zα with other proteins in the database identified as E3 protein and ZBP1 (then known as a mouse expressed sequence tag (EST)). We also identified the winged helix-turn-helix motif. Herbert A, Alfken J, Kim YG, Mian IS, Nishikura K, Rich A. A Z-DNA binding domain present in the human editing enzyme, double-stranded RNA adenosine deaminase. *Proc Natl Acad Sci USA*, 1997;94(16):8421–6. Epub 1997/08/05. PubMed PMID: 9237992; PubMed Central PMCID: 22942.

matching amino acids, we could predict whether the amino acids folded together to form a helix, similar to that which Linus Pauling had famously predicted, or instead form what is called a coil. It seemed that the Zα fold was a helix-turn-helix motif with a wing composed of a β-sheet (Figure 4.8). Since a HTH motif was found in a number of well-known B-DNA-binding proteins, like the globular domain of histone H5, maybe the protein was not truly Z-DNA-specific after all?

The journal *Science* rejected the paper that described the discovery. By mistake, we were copied on a reviewer's comments that were meant for only the editor. The reviewer recommended publication. Barbara Jasny, the editor involved, said "That's not what he meant". I asked Barbara how that could be so. She said that she was not free to disclose the reasons. Another reviewer asked how we knew it was not binding to something else like "half B-DNA and half Z-DNA". The obvious response was "Can you be half pregnant?". None of the results were consistent with that possibility. The only two minimum energy conformations were B-DNA and Z-DNA, whereas unlabeled B-DNA in a 10,000-fold excess did not competitively bind Zα to the labeled Z-DNA probe.

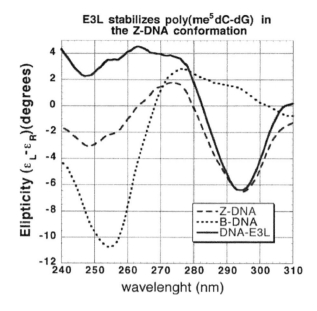

FIGURE 4.9 The Zα domain from the long isoform of vaccinia E3 protein also led to Z-DNA formation but required a methylated polymer and overnight incubation, consistent with later findings that the protein captured Z-DNA, rather than inducing the flip from B-DNA (unpublished). Herbert A, Alfken J, Kim YG, Mian IS, Nishikura K, Rich A. A Z-DNA binding domain present in the human editing enzyme, double-stranded RNA adenosine deaminase. *Proc Natl Acad Sci USA.* 1997;94(16):8421–6. Epub 1997/08/05. PubMed PMID: 9237992; PubMed Central PMCID: 22942.

Alex would not push to have *Science* publish the paper and would not send it to *Nature.* He had not been involved in the work and the first he knew of our success was when I handed him the penultimate draft of the paper for him to review. Why did we not tell him sooner? Ky and I had no doubt in our minds that Alex would start a new lab member on the project without informing us or them of each other's work, just as he did with the crystallographers. He communicated the paper to the *Proceedings of the National Academy of Sciences.* I believe he chose this avenue as publication was almost automatic since he was a member of the Academy. That meant that there was little risk of being beaten to press by some unknown group or having to deal with a reviewer who would likely be one of the two other personalities in the field, each with an agenda of their own. The downside is that everyone knows that the peer review in these cases is not thorough, so the paper is given less attention than it deserves. Also, the junior authors miss out on the kudos of publishing in a top-tier journal.

Meanwhile, I continued the work characterizing the Zα domain. A student, Marcus Schade, and I focused first on mapping the key residues involved in the binding of Zα to Z-DNA by mutagenesis using the Southwestern blot assay that I had developed [44]. Markus seemed not to respond well to the Susumu-inspired chat about "accidents" when we first started working together, but he took the advice in his stride and characterized a large number of Zα variants quickly and efficiently.

I also showed that the interaction of Zα with Z-DNA was structure-specific, not sequence-specific, by demonstrating the binding of Zα to a wide range of different Z-DNA sequences. Again, the approach was super simple. I made probes that had the d(CG)n sequences combined with the test sequence and counted the number of bands present in the assay. If two bands were formed, then both halves of the probe were bound. If there was one band or none, then the sequence was not bound by Zα [45].

We also confirmed the binding of Zα to multiple sequences by atomic force microscopy in a collaboration with Yuri L. Lyubchenko at Arizona State University. Yuri subsequently followed up on the findings with others. Yang-Gyun Kim, who had just joined the lab, fused the Zα domain to the nuclease domain from the restriction endonuclease *Fok*I to make a Z-DNA-specific nuclease [46]. As Z-DNA was resistant to being cut by the enzyme, only the B-DNA either side of the left-handed segment was cleaved. For reasons I do not understand, Yang-Gyun thought he should have been a co-first author on the paper we published describing the discovery of the Zα domain. I don't know what Alex told him, but I do know whatever was said did not resolve the issue. The Z-DNA-specific nuclease was published in a separate paper.

Our focus was also on structural studies of the Zα domain, preferably with Zα bound to DNA. It was exciting when the results obtained by the students, Thomas Schwartz, working with Mark Rould from Carl Pabo's lab [47], and Markus, working with Chris Turner from the engineering department [48], confirmed the features we had predicted based on our biochemical and mutagenesis studies. Their structures were at high resolution and of excellent quality. The crystal structure showed the Zα docked to Z-DNA. The Zα domain was indeed a wing-helix-turn-helix (wHTH), as out bioinformatic analysis suggested. The contacts were as we predicted and the 6-bp Z-DNA helix bound two Zα molecules, a finding consistent with our biochemical studies. The solution studies confirmed that the residues essential to binding were pre-positioned to bind Z-DNA. There was no doubt that Zα bound to Z-DNA in a structure-specific, rather than a sequence-specific, manner.

So, what made Zα specific for Z-DNA when other wHTHs bound to B-DNA? The answer was one we had not guessed. The *syn* conformation of one base in the dinucleotide repeat is the key element for recognition (Figure 4.10). This particular orientation of the base positioned over the sugar is recognized by the conserved tyrosine present in the third helix of Zα. The tyrosine is held in position by a tryptophan, also conserved, in the Zα wing. The Zβ does not bind to Z-DNA as it lacks the conserved tyrosine. Instead, there is an isoleucine present that does not make the specific contacts necessary. This time, *Science* accepted the paper.

It was odd that there was never a laboratory party or any celebration initiated by Alex to mark our success. I am sure he celebrated, but not with us. It had taken me 12 years to go from nothing to finding the first Z-DNA-binding protein, with many of the early years full of failures. Designing an assay with stringent controls was

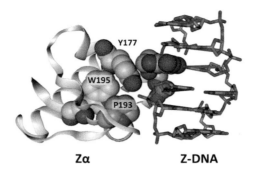

Zα   **Z-DNA**

FIGURE 4.10   Zα bound to Z-DNA. The conserved tyrosine (Y177, colored yellow) contacts the C8 hydrogen of a guanine base (colored pink) in the *syn* conformation. The conserved tryptophan residue (W195) positions Y177 to make contact with the Z-DNA. The proline residue (P193 colored green), along with the asparagine (N173, not shown), have variants that I used to map Zα to disease phenotypes.

the correct strategy, although, at times, I had to wonder. There were certainly many people who thought that the work was a waste of time and told me so in one way or another. Many of them, I suspected, were saying to me what they wished that they could say to Alex, but couldn't or wouldn't mention to his face. Alex also had his doubts. Many of the papers he had published on Z-DNA-binding proteins early on, before I appeared, were, at best, works-in-progress. The results sections were over-flowing with optimistic interpretations, but the required controls were noticeably absent. When new members of the group arrived, Alex would often forget to intro-duce me and Jeff. At some point, we had become just part of the furniture. He was not engaged in our work. Clearly, we were still there doing something. The impact of our experiments on the lab budget was minimal as the source materials were free and the assays were very basic. As long as we did not submit purchase orders of over $100, we could procure what we wanted. There was no need to discuss anything with Alex. Perhaps he just noted to himself that I never missed a day and I just came to work (see chapter 5).

The reaction from others outside the lab to our success in finding a Z-DNA-specific protein was almost zero. Some responses were along the of "I always knew there would be one". The ADAR field just shrugged off the finding as being either irrelevant or an artifact. Ron Emeson, forever articulate, bet on the artifact, but still has not settled the wager.

The RNA editing field in itself was new and still controversial. Brenda Bass had shown that there was an enzyme that destabilized dsRNA. She found that the activity converted adenosine to inosine. Her focus was on an activity that destabilized double-stranded RNA (dsRNA). Only later did the focus shift to recoding of an RNA message. The substitution of inosine with adenosine changed the translation of the RNA, causing production of a protein with a different amino acid at the modified position (Figure 1.7). The obvious question was "Does recoding of protein by RNA by editing have a biological function?". That was the main focus of the field and the work was not progressing well. It was difficult to find examples of recoding.

In principle, editing allowed production of different protein variants without the need to mutate the DNA. The amount of each variant produced could then change the context, with natural selection acting over time to select the best outcome. This idea received initial support when the first examples of recoding were associated with neurological effects. One, in particular, where editing led to the replacement of glutamine with arginine (QR editing) in a glutamate receptor ion channel. If this editing did not happen, then mice would have repeated epileptic seizures due to the increased conductivity of the channel and die soon after birth. The two results crashed the field. First, mice that lacked ADAR1 adenosine-to-inosine editing enzyme activity were normal with no developmental defects. Rather than die as embryos, mice lacking ADAR1 editing activity could be rescued by a second mutation which prevented the induction of an interferon responses by dsRNA, an outcome normally triggered by viral infections. Second, knockout of the related ADAR2 had no phenotype when a mouse had a single adenine base in only one particular glutamate receptor replaced by guanine. The hardwiring of the gene to replicate the effects of glutamine to arginine (QR) recoding was all that was required to compensate for the loss of ADAR2 [49]. The RNA editing field was on the defensive as the biological relevance of recoding was rendered moot by these genetic studies.

The last thing it seemed that people in the RNA editing field wanted was anything to do with something as controversial as Z-DNA. During this phase, the best I could hope for is that those in the RNA editing field included a cartoon in their papers of ADAR1, showing the presence of Z-DNA-binding domains. Usually this was not the case. The only features drawn were the three double-stranded RNA-binding domains and the deaminase domain that performed the editing. It was as if the Z-DNA-binding domains, just like black swans, did not exist. What's worse, if ADAR1 editing was of questionable worth, it seemed that Z-DNA was considered exponentially less important.

There was no traction at all for the Z-DNA part of the story. My response was rather contrarian. I responded: "If you say the Z-DNA-binding domain are irrelevant, how do you know that the dsRNA binding domains are really that important? Do you need those domains for ADAR1 to edit?". That was obviously a dumb question to those involved in the field. The answer was an obvious "yes". In the last paper I published from MIT, I showed that the deaminase domain by itself was sufficient for editing: you didn't need the Z-DNA-binding domain and you certainly didn't need the dsRNA-binding domains (Figure 4.11) [50]. The deaminase domain defined the residues that were edited. The two different types of Z-DNA and dsRNA structure-specific domains were there either to localize the enzyme to a substrate or to alter the kinetics of binding to the substrate. I supported that role by demonstrating that mutations to tryptophan 195 in the Z-DNA-binding domain decreased the editing percentage of a subset of short dsRNA substrates I examined by 28% even though they lacked Z-RNA-forming sequences. The result suggesting that the Z-DNA formed during transcription from the plasmid was playing a role in localizing ADAR1 to the RNA. Zα was doing something. In that paper, I also determined that a minimal editing substrate had a 12-bp dsRNA stem. Others followed a few years later to make constructs using the editing domain alone to recode Mendelian disease variants in order to create an error-free messenger RNA. The work based

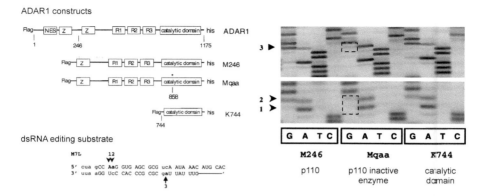

FIGURE 4.11 Comparison of the editing of wild-type and inactive ADAR1 p110 and the catalytic domain-only construct. The assay at that time was performed using Sanger sequencing, with each lane giving the relative position of each nucleotide in the base ladder. The arrowheads indicate editing of an adenosine to give a guanosine residue, with the dotted black box showing the lack of editing with a catalytically dead enzyme. The numbering 1,2, and 3 correspond to the positions on the dsRNA editing substrate. The asterisk in the Mqaa domain shows the site of the loss-of-function mutation. Herbert A, Rich A. The role of binding domains for dsRNA and Z-DNA in the in vivo editing of minimal substrates by ADAR1. *Proc Natl Acad Sci USA*, 2001;98(21):12132–7. Epub 2001/10/11. doi: 10.1073/pnas.211419898. PubMed PMID: 11593027; PubMed Central PMCID: 59780.

on these minimal substrates also supported the use of short editing guides to guide recoding. This therapeutic approach is now rapidly progressing to the clinic, as we will discuss later.

It was, however, time to leave Alex to his own devices. There were various issues that had arisen. During the period I worked with Alex, during the dark days when there was nothing to say about Z-DNA-binding proteins, a constant stream of talented post-docs came through the lab looking to work on Z-biology. They were pretty much left to their own devices with infrequent group meetings. For me, I had finished what I had come to do. It was good to arrive at an answer and time to move on as my success did not translate into a better future. Staying in Boston had worked out well for my children, who benefited from the local schools. By now, two made it to Harvard and one to Yale. At the time, no one seemed that interested in hiring someone working in a field like Z-DNA that no one thought worth pursuing. Alex did not help with securing interviews at any of the places I applied for. To be fair, there was a huge stretch where I had no publications, even though there were seven in 1999. I was competing against people who were at an earlier stage of their careers with first-author publications in *Cell*, *Nature*, or *Science* and strong supporting letters.

After a rather forthright discussion with Alex about the situation where I pointed out to Alex that "high risk, high reward" research was not about me taking all the risks and him taking all the reward, there was no doubt about the need to move on. Alex must have received a phone call or glanced at a job ad that turned up in his mail because in no time flat he had proposed me for a job at Boston University running the genetics lab for the Framingham Heart Study. Just a phone call was all it took. Surprisingly, the complete change of field set off a series of events that would lead me back to Z-DNA almost 20 years later with the skills needed to prove the biological relevance of left-handed Z-DNA and Z-RNA and to make a few more unexpected discoveries.

Alex died in 2015 not knowing the biological function of Z-DNA. I often wonder about what would have transpired if Alex had followed Crick's advice about his junior colleagues. The letter dated September 4, 1974 (Figure 4.12) was part of the correspondence relating to the tRNA controversy, and perhaps referred to the breakdown in communication between Alex and Sung-Hou.

Would Alex have celebrated with us if he had taken better care of those many young aspiring scientists who placed their trust in him? (Figure 4.13)

I think you would be wise if in future you took especial care that you respect other people's priority and, what is equally important, are clearly seen to be respecting it. You already have an established scientific reputation and I think you should try to bend over backwards to acknowledge the ideas and influence of others, not only about work in other laboratories but also by junior people in your own. I know from personal experience how very import-

FIGURE 4.12   Crick's letter to Rich (http://resource.nlm.nih.gov/101584582X216).

FIGURE 4.13   Picture of Francis Crick, Alexander Rich, and the author taken by Shuguang Zhang.

5 Failing Successfully Everywhere Else: The In-between Years

Early on, I was often questioned as to why I worked on left-handed DNA when there was not much evidence for a biological function for this conformation. The whole world was working only with the right-handed Watson-Crick model of DNA, but not me. I was told that the whole army can't be out of step so it was me that needed to change. To fast forward, there is a biological function for the left-handed helix. In fact, there are many. The different conformations of DNA offer a different way to encode genetic information. Looking back, the whole army was unable to imagine anything different from what they had carried forward from the past. Tradition, as the song from the Broadway hit show "Fiddler on the Roof" describes, makes change difficult!

By nature, I ask a lot of "What if" questions. People initially find the approach interesting and tolerate it for a while. Over time, it becomes tiresome and, at some point, the answers run dry. This lesson is one I should have learnt early when, as a medical student, I signed on as a New Zealand Army volunteer. We all mostly joined to gain some extra money – what student doesn't need that – and the time requirements were rather modest. They did not interfere with the other jobs I took to finance my education. However, it did occur to me during the drudgery of practice drills, that there might be the need for a few updates to the army protocols specifying how soldiers should march. No, not a change to the left, right, left, right part. It seemed to me that the problem was elsewhere. The regulation step was established when the average height of the enlisted man was rather short by modern standards. Maybe it was designed with child soldiers in mind? What was natural for that fellow back then meant anyone much taller right now was short-stepping all the time. What if, instead, we increased the regulation step size to match how real men stride out? Wouldn't that be better as we could advance the troops into battle faster! No surprise, the question was not well received and the answer was "no". Choice is not an option in the military. You do what you are told and there is no need for questions. That reality of the routine became more apparent to me later when, after five days of rifle drill, I mastered the art of advancing the weapon from the position of "At ease", one movement at a time, to the stance of "Take aim". A new day, a new motion, another step in the progression, practiced over and over again. On the final day, I realized that, with the rifle at my shoulder and cocked to fire, I would, by reflex, pull the trigger on command. There was no thought of who or what was aimed at, let alone any need to question why. I had joined the Army to save lives, not end them. On day five, there

DOI: 10.1201/9781003463535-6

was no thought of that. While in awe of the precision of military training, I realized that it was not for me. The military also came to the conclusion that I was not for them. They discharged me, allowing me to keep my boots. Army boots are great for recreational hiking, where you take any length of step you like.

One curious thing about the military is that outstanding performance is usually rewarded with a medal awarded posthumously. You likely will never get to know which of your contributions on the battlefield was really outstanding. On the other hand, the church has solved this particular problem. It guarantees that you will receive your reward in the afterlife. Cleverly, your recompense will be one of those eternal mysteries that others are left to ponder. My father, who, among other things, once sold insurance, liked to tease his brother-in-law, who, as a member of the clergy, was clearly closer to God than his less-educated brother-in-law could ever be. "You know Rex" (note that in Latin Rex means king or ruler), my father would say, "we are both in the same business. There is only one difference. I sell fire insurance to someone while they are still alive and you sell it to them for after they are dead. Your business is clearly more profitable as you never pay out".

I also learned the power of the "What if" question when, at the age of 13, my father thought I could take my brother's job at the local abattoir. My brother did not like the job: it was menial, just threading string loops through small cardboard tags – similar to those you attach to luggage. The tags were destined to tour the world, but not those workers performing this one-step assembly task. The tags were attached to the fully dressed lambs at the end of the chain before being sent to cold storage. Even then, an inspector assessed the quality, awarding some of them top grades in their afterlife.

I don't blame my brother for his decision to leave as the job was without redemption, apart from the pay. Even the supervisor assigned to overlook the bored taggers was bored. His job was to ensure that everyone tied 3,000 tags per day. For me, he provided the ideal incentive. What if the target was reached, I asked, "Could I leave and receive credit for the whole day?". "Why, yes you can!" was the reply. On most days, I started at 7 am and was gone by noon. Job and finish, it was called: a philosophy I still use to structure my day.

Over the next few years, during vacations and while still at high school then college, I continued to work at the abattoir. I used to get there by train, then by a 50cc Yamaha motorcycle, then a dirt bike, then a 500cc surplus police bike I picked up at auction with everything on it except the flashing lights, then a 750cc Yamaha with a shaft drive when I became tired of being pulled over because of the former police bike. I switched to a job in the freezers when I was 16. Since New Zealand is far from the market it serves, the meat was exported by shipping the frozen goods in containers. The work was well-paid and was considered one of the prime jobs at the establishment. No surprise, but it was critical that the labor met the shipping schedule.

The bosses wore white overalls with spiffy blue trim. Their names were even embroidered in blue above the left breast pocket. The bosses were known as the "white coats". To keep the workers happy, the management made many concessions. Most importantly, they kept the union representatives happy since those guys managed the other side of the line. They took care of problems that the white coats couldn't. I don't know the *quid pro quo* involved, but everyone in the blue denim had a grateful story to tell how the union guys had helped them out of a jam in the past. Usually everything went smoothly, and both sides of the table maintained a respectful distance, often turning a blind eye to the petty theft of meat and a few other minor transactions that weren't strictly by the book. I never had problems with either the white coats or the union guys. I would just show up at the start of my vacation and find a white coat to sign me on and usually would start that day. The white coats liked me as I never missed a day and I just came to work. The union guys liked me as I never missed a day and I just came to work.

The white coats were clearly not one of us, as we wore blue denim guaranteed as good down to minus 30 degrees Celsius. The white coats would come into the freezer to direct us where to go and what to do when we got there. Then, they would wait outside. At ten minutes before the hour, they would come in and say it was "smoko". That meant we could warm up outside for ten minutes, lost in our thoughts.

The freezers were four stories high. To move the meat from top to bottom, someone was placed in the hole – the opening between the freezer floors. They stood on a platform and grabbed the frozen carcasses tossed there by other members of the gang. The person in the hole would then place the blocks of meat on a slide that would eventually take the goods all the way down to the loading dock. The white coats were careful who they placed in the hole. You didn't want to be in that place if one of the gang was at odds with you. For reasons of safety, the white coats would also cull certain individuals into different gangs. Noel and "Butch" were two individuals with whom one had to be careful. After Noel took a strong dislike to me because I was taking a job away from someone who needed it more than me, or because he was a Vietnam veteran and I was one of those entitled trouble-making students he disdained, we were never placed on the same gang. Noel was also the one sent home early when too many hands showed up for overtime. Nobody wanted us both clocking out at the same time when there weren't too many other people around. Nor were Butch and "The Colonel" ever put together. Even in the changing room, their storage lockers were at opposite ends of the shed. Neither would cede ground to the other. The Colonel was then in his sixties and in good shape. He had served in the British special forces, but never felt the need to prove himself. But Butch did, and the more Butch did, the more the Colonel stared him down; never a blink, feet astride, arms poised, just waiting for Butch to bring it on; that never happened. Most other issues were settled off-premise.

We had one hour for lunch. That was enough time for some to go to a local pub and swill a pitcher of beer. Many others just napped on the locker room floor as we mostly worked 12-hour shifts and six hours on Saturday. The work was tiring. There is an art to tossing a 60 lb. carton of meat on top of a 7-foot stack. Removing the steel gambrels that set the hind legs of the sheep straight as they froze was also an acquired skill. There was a knack to removing the gambrel and stacking the frozen carcass all in one movement.

The pay was great, especially with all the overtime. There were a few high-stakes card games played every Thursday. Before lunch, everyone would assemble near the pay office to collect their earnings in envelopes stuffed with cash. Then, back to the lunch room where the cards would come out. No one wanted to play with Butch but then nobody wanted to say no when he sat down to join their game. The types of games I saw were never worth playing. They were not friendly. Everyone seemed to know that everyone else was cheating, but nevertheless, they knew they were the better cheat. The jousting made up for the every day drudgery of the job.

Of course, every once and a while, it was necessary for the union delegates to remind the white coats never to cross the line. Without much of a preamble, we would hear the word that the white coats had gone too far. We would drop everything, literally, and be headed back to the shed in less than ten minutes. It was called a walk off; the army never moved so quickly. We were on strike within the hour. It would have been sooner, but we had to count the votes to show solidarity. Democracy in action, union style. The entire plant would grind to a halt. The white coats were left to bag the meat, clear the cooling floor, and stack the carcasses in the freezers without us. That showed them!

I never knew why we went on strike, except for one time. The white coats took exception when a whole container load of meat was lifted off a railway wagon and placed on a truck that seemed to vanish without anyone seeing anything. The white coats locked us out for a week. The union leaders took us out for a month over the unfairness of it all. That really let those white coats who was boss! If those white coats brought in replacements, then every other plant in the area would close down in support of our action. This time, that did not happen, nor did any more containers magically disappear once we returned. Lesson learned. We all needed our pay checks. No one was criminally charged for this transgression that I knew about, but there were two forklift drivers who did not return to the job. Normally, everything ran smoothly and the work was finished ahead of time. That meant an early finish with all the promised hours paid in full, allowing the boat to leave on the outgoing tide. The white coats were happy; so were the union delegates. We were happy as the line between them and us had not moved, nor had it been crossed and we had money in our pockets.

As I write this, I realize that the white coats and the lines that should not be crossed are a part of many stories. The job and finish, the long hours of heavy lifting, the blue denim instinct for survival, the walk-off, and the tendency for people to redraw the line as it suits them. There were many lines that criss-crossed in my story that created a tangled web full of enticements and entrapments, but most were best side-stepped, especially when the odds were uneven or clearly doctored. .

Between bouts of work at the abattoir, I went to Medical School. Like the English system, entry in New Zealand was directly after graduation from high school. The Medical School experience lasted for six years with another year to become fully registered, after which you were free to practice the art of medicine. I had never wanted to be anything other than a doctor from as early as I can remember. Probably that was my father's wish more than mine. My father was a dairy farmer, working land that he had converted from a swamp into productive pasture. He acquired the property soon after the end of World War 2 when it was awarded by lottery. His father had died when he was 16, taking away his chance for a college education. He did have one semester before being called home after my grandfather's premature death to support his mother Annie and his sister Audrey. Annie had been sent out to New Zealand at the age of 16, unaccompanied, by ship from County Cavan in Ireland, to be with relatives, who were already settled farmers. She was Protestant and never grew beyond her roots, always warning me about those Catholic girls. Apparently, she knew a few back home in Cootehill and had heard stories about how they behaved.

My father soon tired of the farmer's life. He married my mother when she was 28, old for that era. At the time, she was working in the post office. My mother had grown up as a daughter of a well-off merchant who had stores in a number of towns that sold provisions to the farmers. That side of the family was sold land by one of the companies set up to settle New Zealand in the 1800s on the assumption of vacant possession. The Maoris, who arrived a few centuries earlier, objected to this arrangement and a war soon followed. Stone age versus iron age: not a fair fight. My mother's brothers became lawyers and stockbrokers, except for the one who became an Anglican minister. Apparently, that is the English tradition when more than one son is standing in line to collect an inheritance; one gets the farm, others are sent to the colonies, and the fool of the family joins the Church. My mother had no college education. After marriage, her life changed suddenly. From the social life of a small town, she found herself isolated on a farm with three children under three. By the time I was five, my father moved me, my older brother, sister, and mother to the city, ostensibly to further our education. That meant moving 80 miles north to Auckland, the largest city in New Zealand. He kept the farm and operated it in partnership with a share milker, who owned a dairy herd, but not any land to graze them on. The profit from the sale of milk was shared equally between the partners. My father kept himself amused in Auckland by selling agricultural chemicals, real estate, and insurance, meaning he was not around all that much, just enough to add two more brothers to the family. When he was home, we were doing renovation to houses we lived in – usually jacking them up high to add floors underneath or digging below to add basements or pushing them out sideways to add rooms. That is the need when you have a large family – that and a big car and a trailer to pack everything into when camping is the only affordable vacation you can take. He had a cast of characters who would come around and do plumbing, roofs, and staircases, each of whom it seems Dad had helped out in some way or another. There seemed to be more trading in kind than cash involved.

I used some of the money from the abattoir to buy my first house at 16, and paid the bills by renting out rooms. The house was a "fixer-upper". It was built on timber pilings that had rotted and even the brick chimney had sunk. The front of the house was 16 inches lower than the back. Of course, my dad had a few friends who could jack everything up to bring the floor back to level. My father would always have some guys ready to step in if help were needed.

My father did ensure that we received a great education. My mother did her best and would help with homework, until she couldn't. That was when I was about 10. After that, it was beyond her. I ended up going to Auckland Boys Grammar. It was a single-sex school and the best public school in New Zealand. My name is on the Honors Board for winning a national examination scholarship that helped with my admission to medical school. It was also pretty much a job-and-finish situation, with exams requiring little preparation. I was in what was called the Latin stream which was for training future lawyers and doctors, while my brother was in the technical stream for training mechanics and tradesmen. I did well in sciences and was head laboratory boy – tasked with helping set up science experiments for classes. My union experience came in handy as we were able to negotiate benefits equivalent to the prefects who were selected as role models for the junior classes, and those players in the first XV rugby team and first XI cricket team. This position led to my selection in my final year at high school to represent Auckland Grammar at the Edison Conference for Young Scientists in Melbourne, Australia. That was the first time I boarded an airplane – the scenery, the conference, and the science inspired me to reach further.

At medical school, I cruised my first year and then pulled back in my second year after discovering the opposite sex was a lot of fun to spend time with. Also, in the second year, I had to deal with continual assessment for the first time. Rather than just a final exam, which was pretty much job-and-finish, weekly assignments and frequent tests were the major determinants of grades. There was not enough time for everything so "What if" I spent less time on that assignment in psychology and did my own experimentation with the college-age students of my choice. Why live vicariously? I found my approach more fun than reading about other people's studies performed on the same subject pool!

The other downfall in that year came when I actually started reading books – prior to that, I would just sample enough of an assigned text to find sufficient, suitable quotes for my English teacher to grade my essays above fail. I can still recall the quotes but don't ask me what the novels were about. Like any random process, I occasionally scored an "A" for my original interpretation, along with a few fails for completely missing the point of the piece. Telling Mr. Bone, the head English master, that his questions were often poorly formulated because it was not clear what he was asking certainly, did not improve matters.

Books by Jacques Monod, François Jacob, Julius Axelrod, John Eccles, Albert Szent-Gyorgyi, and many others from the Penguin Collection in the Science Section of the University Book Store became a constant source of inspiration and enlightenment. All full of new insights for me and finally something really exciting to read!

As medical school continued, my situation only became worse as scheduled lectures and labs grew to 40 hours per week. There was no time to question. Here, rote learning was the order of the day. You learned the rules of thumb that worked, given the particular bias of the assigned grader. Failing that, if you were short of a suitable answer, acting confidently was an adequate response. No time for "What-ifs". You had to act. It was soon noticed that I was not in step.

Being on that list was rather unfortunate as it did cause a few extra problems. Only at this time it was not called a PIP (personal improvement program). On one occasion, I was called to account by the Dean for some misdeed that I was unaware I had committed. Thankfully, the misdeed was not mine and definitely not the one that I thought might have necessitated a heart-to-heart with the Dean. Clearly, the issue was "Who else but Herbert would have done such a thing?". It seemed that I was the first name that came to his mind. Only later did I receive an apology from the Dean for this case of mistaken identity.

On another occasion, they had the name right, but the wrong person. Herbert was summoned to an operating theater where a patient who Herbert had clerked was on the table. Herbert was then roundly made aware of his tardiness and other shortcomings in some rather crude language, so I was told. The reason I don't know exactly what was said is that the Herbert involved was not me, but rather the Vice-Chair of the Department of Obstetrics and Gynecology, Herbert Green. That Herbert was on the receiving end of the tirade. Apparently, the senior surgical resident had not looked up to see who he was addressing in such dulcet tones. Fortunately for me, I had excused myself from attending the day before with the junior surgical resident, but the message apparently was not passed along. Unfortunately for me, as a reward for my bad form in not being there, instead of Herbert Green, I was failed on that course and invited to take a make-up exam. The intent was to let me know that such a situation should never happen again. I fully agreed with that notion.

By that stage, a career for me in science was pretty much set and supported by the faculty as the best resolution for all. In fact, I was advised that the only reason for me to graduate was that scientists with a medical degree are paid more than Doctors of Philosophy. So, with that decided, it was time to take a short break. I was 22 years old and married shortly afterwards. Our honeymoon lasted nine months. It was spent driving around Europe in a VW Kombi van ticking off in our green Michelin guide visits to various art galleries, churches, and piles of historic rubble along the way. We went to see all the highlights from those history lessons at school intended to celebrate our status as a colony of the great British Empire. The New Zealand school curriculum featured the ancient and hallowed wisdom of Greek and Roman scholars, enhanced with a touch of Shakespeare and the English television programming shown in black and white on the single Government-run broadcasting channel we had. Of course, there was rugby and cricket for the boys and field hockey for the girls. We were all New Zealanders with a single culture based on the hallowed English traditions.

The trip through Europe was largely financed by our very friendly bank manager, using as collateral the house I purchased when I was 16 and that had rapidly increased in value as the location had become yuppified. At that time, currency exchange was strictly controlled by the New Zealand Government. We were able to draw $250 per month through our American Express Card. Each month, at some place in Europe, we would appear at their establishment and make the withdrawal. Combined with sleeping in our van and armed with a Michelin Guide that also listed when admission to historical sites was free, we stayed on-budget. To find affordable food, we would carefully follow the lead of a local matriarch in her knee-length, long-sleeved coat with matching hat and empty shopping basket. While we did receive some suspicious looks from our quarry, we were never hungry. Our van held up even though the tires kept blowing out, the brakes never properly worked and the sliding door fell off in Venice. The trip ended when our bank manager sent a note delivered by way of the American Express Office in London. He informed us that we had spent enough and it was time to come home and pay it all off. It also turned out the reason Penny stopped drinking Sangria in Spain was that she was pregnant. On return, after raising a third mortgage and doing some locums to pay off bills, I started my PhD at the University of Auckland in the Department of Pathology.

Obtaining degrees one after another, rather than concurrently, was the way things were done back then. My father took it in his stride that his son was no longer going to be a medical doctor just as he realized none of the other three sons would ever become a farmer. He said he understood. He would always handle such setbacks with humor. He told me of a farmer he knew who would be out in his front paddock day after day, just looking into the sky. Finally, curiosity was such that eventually someone spoke up and asked him what he was doing. The farmer said that he read about how to be awarded the Nobel Prize. First, you had to come up with some good ideas and then be out standing in the field. So, he had come up with some really good stuff and was waiting for those people to drop by with his medal. That's New Zealand humor for you...(and a modicum of truth to it - see Chapter 9).

My degree program was in the newly established Department of Immunobiology headed by the other James D. Watson, not the one that proposed the model of DNA along with Francis Crick. Fortunately, Jim had just returned from the States. He was an immunologist who had shown that the response to a certain type of bacterial wall component was controlled by a single gene, eventually shown to encode toll-like receptor 4. He had also worked with Steven Gillis to purify one of the first signaling molecules that drove immune responses, namely Interleukin-2. Jim returned to New Zealand, while Steve went on to form Immunex. I don't think Jim realized that there would be such a huge difference in financial outcome. Immunex will feature in a later chapter.

The laboratory focus was on immunology and I published some papers on the population of cells, called natural killer cells, that attacked tumor cells after being stimulated with Interleukin-2. This type of lymphokine-activated killer cell was used in cancer therapy for a while by Steven Rosenberg at the NIH, but was of limited utility because of its toxicity. Jim was able to persuade a number of prominent US scientists he had met to visit New Zealand, including David Baltimore, Alice Huang, Wally Gilbert, Hugh McDevitt, Ave Mitchison, Dick Dutton, and Suzie Swain, allowing opportunities for one-on-one conversations with them all. The experience was great for any graduate student. David and Jim had been at the Salk together, along with Susumu Tonegawa, who we will meet later.

I was not sure that Jim was ready for the clash of scientific cultures when he arrived back in Auckland. Jim returned to an environment where the debate between epidemiologists and biologists was in full swing. It was focused on the best return on investment for a country like New Zealand on research dollars expended. The epidemiologists argued that epidemiological studies had led to public health and life-style choice interventions that were responsible for the remarkable gain in life expectancy. They referenced the decrease in things like smoking-associated morbidity and mortality. On the other hand, New Zealand's Dr. William Lilley had received worldwide recognition for his work on Rhesus factor and hemolytic disease of the newborn while Dr. Mont Liggins had introduced the use of steroids to improve survival of premature babies. The Australian Sir John Eccles also had won his Nobel Prize for work performed partly in New Zealand, when he was head of Otago's Department of Physiology. He invented a device to show the transmission of information at neuronal synapses was chemical rather than electrical.

With the dawn of the recombinant DNA era and the ability to ask questions never before possible, the opportunity was there to develop molecular biology further in New Zealand. Once back home, all Jim could do was watch as the biotech industry in the US took large strides and his former graduate Steve Hedrick went on to clone the T cell antigen receptor with Mark Davis at Sanford. Jim eventually put behind him the battles at the University to form the first New Zealand biotech called Genesis, remaining active in shaping New Zealand science policy as President of The Royal Society of New Zealand, a member of the government's Growth and Innovation Advisory Board, and a trustee of the Malaghan Institute of Medical Research.

Jim was really helpful in writing letters on my behalf to secure a post-doctoral position in the States (Figure 5.1). I had offers from Jack Strominger at Harvard and Alex Rich at the Massachusetts Institute of Technology (MIT). Alex and Jack actually had both been roommates while at medical school. I chose Alex. Why Alex? Well, Jack wrote that a previous New Zealander he had had in his lab had not impressed him. In contrast, the mystery of Z-DNA, discovered in Alex's lab, had just been featured in *Science* by Gina Kolata. It was apparent that the opportunities to do such blue-sky research in New Zealand would always be limited, so why not see where that adventure might take me? So, it was an easy decision to go from New Zealand to the new Z-land (with "Z" pronounced as "Zee" in the US).

With Jim's help and a letter from Alex describing all the wonderful discoveries his lab was making in Z-DNA biology, I headed to MIT, financed by a Fogarty Fellowship from the NIH. Jim was very supportive in this process. The move to the United States and Boston in particular was much to the chagrin of my wife, who would have rather gone anywhere in England, which was the motherland for many Kiwis of such stout British stock. No one she knew in New Zealand had ever heard of MIT. All the knowledge she had of the US came from imported television shows.

FIGURE 5.1   The other James D. Watson who I trained with in New Zealand.

Those offerings clearly did not reflect the diversity of the US culture. Eventually, she and our three children joined me in Boston after spending some time with friends in Germany, but reluctantly, and only after her use of our American Express Credit Card was suspended by me. It was not easy financially in those years as my stipend was taxed and barely covered the rent for our house. There was little left over to cover living expenses and we only survived by obtaining permission for my wife to work. She ran a daycare for children who were the same ages as our three little ones. Nevertheless, our family flourished, with two children graduating from Harvard and the third from Yale. The latter son had to endure for each of his four years in college the defeat of Yale by Harvard in their annual football game. That event did not go unnoticed by his brother and sister. So, despite the underprivilege in pay that comes with the privilege of being part of the MIT. community, our children were given a unique opportunity to craft their own life stories.

Throughout this journey, there were many choices I made about which battles were worth fighting. These decisions involved a lot of "What if?" and "How about?" questions. Usually, after working through the data and lots of scenarios, a clear course of action emerged that allowed for a job-and-finish outcome. In science, this means asking a question that delivers a clear "Yes" or "No". It often takes much more of your time if you do a bad experiment than if you stop to design the best one you can. The challenge is to find a way to simplify the problem sufficiently to give an unambiguous answer. It requires that you focus on primary data. Statistics will only give you a threshold of significance and often those based on small numbers give a biased estimate. Everyone likes to confirm their favorite hypothesis, as is evident in many studies that fail to reproduce, so they grab the numbers with the best fit.

A life in science is challenging in a number of ways, both personally and for those around you. I think the fear of failing is difficult to manage, especially when the experiments do not work as you hoped and when more money is going out the door than is coming in. A lot of anxious moments! On the other hand, as you work through the different scenarios, it can be difficult for others to know where you stand and difficult for them to understand the route by which you arrive at the decisions you make. At one point, you may be totally convinced that you are on the right track, only to soon discover that your approach, in fact, is fatally flawed. The cycle progresses with each new hypothesis, right or wrong. To the observer, it all seems impossible, but then, if you are fortunate, it is not. My former wife gave up trying and just told her friends that I worked on cancer, even though this was not true even 15 years after our divorce.

Many others along the way have reached exasperation with my approach to finding answers. My initial meeting with my medical team following my diagnosis with stage 4 squamous cell carcinoma of the throat reached that point rather quickly. After about 50 minutes with my medical team and after asking many "What if?" and "How about?" type of questions, one of the physicians said there is only one treatment they offer and that it had been explained to me in detail. I replied that he clearly has these sessions many times a year and I was doing this for only and maybe once in a lifetime, so shouldn't it be alright with him that I asked so many questions? In the end, I agreed to his suggested treatment and thankfully that was the correct choice!

There were many other challenges before and after the bout with cancer. During the MIT years, I just kept plugging on, trying to do my job-and-finish act. I thought the question about Z-DNA was worth answering. As prospects for a biological role for Z-DNA dimmed, the only students Alex could attract were from overseas, predominantly German. The MIT experience added to their curriculum vitae and was accommodated by the milestones academics in Germany must achieve by a certain age in order to advance their career. Overall, they were a talented group of individuals with both moral and financial support from their home institutions. As a whole, the students had far better career outcomes than Alex's post-docs of that era.

The way Alex managed the interactions could make it difficult for the students to navigate the waters in the lab. In one instance, it was rather surprising for me to pick up from the communal printer in the lab, what I thought was my print job. Instead, it was a sheet from one of the students I was working with. I saw my name midway down the page and was momentarily confused. The printout was of a protocol using "Alan's reagents" to perform an experiment with another post-doc in the lab, someone who didn't want to collaborate with me. Apparently, the student was working with the other post-doc under Alex's direction. This time it was not about "borrowed data", but about "borrowed reagents".

There were other instances like this, where there was an attempt by a student to rename the Z-DNA domain I had called $Z\alpha$ because he constructed a version that differed in length from the one that I cloned. The domain the student worked with was equivalent biologically but he thought his was better for crystallography. These and other incidents that did not involve me just said to me that it was time to do something different. I had found the Z-DNA binding protein I had sought, along with others that were related. At the time, there didn't seem to be much opportunity to work further on Z-DNA in my own laboratory. No one seemed to care about how challenging the task was of finding a Z-DNA binding protein from normal tissues has been or thought it was worth taking the work further. The key papers from that era are not often cited even though my work 20 years later leaves no doubt as to the biological significance of the $Z\alpha$ domain. Editors limit the number of references allowed and also prefer those that are the most recent. Of the more than 4,500 papers found by searching Google Scholar with *"Zα"* and *"Editing"*, just over 400 reference the original paper and about the same number reference the crystal structure of $Z\alpha$ bound to Z-DNA. Of course, it is possible that reading papers from the last century is not the "done" thing. My only response is to note the following. It is actually quite interesting the things you can find if you bother to look.

THE FRAMINGHAM YEARS

Leaving MIT, it was an exciting time to start something new. The human genome had just been sequenced. The Framingham Heart Study (FHS) was famous for its role in identifying risk factors for heart disease. I knew it from my medical school training and from the debates in New Zealand as whether research dollars were best invested in epidemiology rather than molecular biology. FHS had many phenotypes collected over two generations of families, all meticulously recorded at a single location. Participants were from the City of Framingham, close to Boston, and, at the time the study started, was representative of the American population demographics. Many worked at the General Motors Plant. The study was one that was prospective. A number of hypotheses had been advanced to explain why the incidence of heart disease was rising. Was it cholesterol, blood pressure, age, weight, ECG abnormality, hemoglobin levels, or the number of cigarettes smoked? The term "risk factor" arose from the analysis, allowing an assessment of the impact of each on the rate of heart attack and stroke during the observation period. The study was deemed so successful in achieving its goals that the NIH decided there was no point funding it further. Everything proposed had been done!

With the help of Richard Nixon, sponsors like the Tobacco Research Institute and the Oscar Mayer Company, and money from private donors, the study survived until NIH funding was restored through an administrative contract with Boston University. There was now money to recruit the second generation. The original idea was to follow trends in lifestyle and other environmental events that would increase or lower the risk of heart disease. For example, do children of the original participants smoke cigarettes more or less? Are their health outcomes different? What is the effect of dietary changes? Do new technologies allow heart disease to be detected earlier? All valid questions.

At this stage, FHS was not conceived as a study to determine genetic effects. Rick Myers initiated the change in emphasis. He started collecting the DNA. He extracted it from old blood samples. Where he could, he asked participants for permission to make permanent cell lines from their blood to ensure an ample supply of DNA for future studies. Initially, the investigators wanted the DNA to find gene variants that caused heart disease. At that time, it was known that different APOE alleles had an impact on cholesterol levels. Were there other variants that increased the risk of stroke? Rick Myers then had an "issue" with Phil Wolf, the FHS principal investigator and left the study. It was around the time I was hired. Within six months, Phil had increased my salary to double what I was earning at MIT. I was very thankful.

Apparently, I was hired because the National Heart Lung and Blood Institute (NHLBI) wanted someone who had hands-on experience with DNA. Maybe it was a good thing that they didn't know the difference between right- and left-handed DNA. My hope was that we could eventually find some outcomes that were related to ADAR1 that might show a trait depend on its Z-DNA binding. Plus, the human genome had just been sequenced. With that data came maps of genetic variations that could be used to find those that affected different human traits. It became possible to perform human genetics in a very ethical way by observing nature at work, taking data collected prospectively, and explaining outcomes in terms of the genes involved. An exciting time to sign on!

I was hired and the next phase in my career started. This experience was my first in the field of political science. Worse than that, I never knew that there was a field that could be called political science. The politics were tied to the funding and to the public perception that great advances were the result of this funding. As FHS was a poster child for the amazing deeds enabled by the NHLBI and for why it deserved a special status when it came to allocation of Federal research dollars, messaging was strictly controlled. All the scientific papers from FHS required review by NHLBI before they could be submitted for publication as an official FHS publication. Ostensibly, this was to check whether the paper complied with various FHS guidelines and to catch any errors that might have escaped the authors' notice. For all their work in conducting the study and performing the examinations, the *quid pro quo* was that FHS investigators had exclusive access to the data they accumulated for a two-year period prior to it becoming accessible to non-study investigators. The FHS investigators could also use the data to apply for additional funding through the normal granting mechanism to explore other hypotheses not funded by the contract. Of course, non-study investigators were handicapped because they did not understand the data structure. What this meant really was that those external investigators had to work with the FHS principals to unlock the information relevant to their question. As Peter Wilson, one of the FHS physicians, said to me, "It's like real estate. An agent does not collect a commission unless they show you the property".

It was even more complicated than that. Each FHS contract was reviewed by investigators from other NHLBI-funded epidemiological studies. You would expect that such a situation would cause competition – may the best study be funded. However, it was more of a game. What was said in one review would bounce back when the reviewer's study was next up for evaluation. It's not too dissimilar from how airlines negotiate with each other for outcomes where everyone is better off. For example, airlines act within the law by publicly disclosing their intent to raise passenger fares. If other airlines respond in kind, then the price hike stands If the other airlines don't, then the original airline rolls out the public relations crew to assure the public that they heard their concerns about a price increase, so refrained from taking that action. You need a few of these apparent disagreements to assure everyone that the marketplace is competitive. Similarly, what could happen to one NLHBI study could happen to another. Given its prominence, the thinking was, as goes FHS, so go the others.

There was a very public discussion of what the next FHS contract should contain. There were other subtexts as well. NHLBI had intramural members assigned to the study. Even though they were housed locally in the Boston area, they were NIH employees. At the time I joined, the NHLBI contingent included Dan Levy and Chris O'Donnell, among many others. The NHLBI folks did not have a budget of their own, so, to do something, they needed to trade the asset. Phil used to joke that when he shook hands of one of them in particular, he used to count his fingers to make sure none were missing. He didn't want me to write grants. Being an assistant professor, I had the right to do so. Fortunately, the Chair of Neurology, Steve Fink, not a member of FHS, took my side. Yes, the FHS was administered through the Department of Neurology because of the work on stroke. So, I wrote grants. There was no assistance provided. Phil would come out of his office and watch me xeroxing the required three copies of a proposal. He left me to attend to the administrative details that he would have his staff do for him. No assistance was offered. Unfortunately, Steve, still a young man, died of glioblastoma soon after I joined.

As a result of Phil's leadership, the Study was very territorial. Everyone had their own turf. Phil had helped keep the FHS going when NHLBI funding lapsed in 1969. So, he ran a tight ship, ever vigilant for any loose cannons. Apparently, I was one, and it didn't take long for this to become an issue. Everyone had their area of investigation with well-defined boundaries. I learned how strong the enforcement was when the genetics laboratory dared to analyze a set of gene variants across all the available traits in the Study. Why not? Genetics is a very powerful tool for the discovery of how genes impact many disease outcomes. After presenting the results to the Framingham investigators at the Framingham site, there were only a few questions. Maybe I had presented the analysis badly or the subject was new to the audience? I was then asked after the meeting to immediately join the investigators from the study in a private discussion. The first question was "Who gave you permission to work on my phenotype?" – well, there was at least one four-letter word in that sentence as in "Who the …. ". Faces were very red. I had not asked for permission! If I had known about and then paid more attention to Alex's experience with the RNA tie club, I would have been better prepared!

The situation only became worse from there on in. I crossed another line by writing a proposal to perform a genome-wide study for disease-causing variants in the FHS population. This section was added to the contract renewal as an appendix to the response to the Request for Proposal NHI-HC-01-2. It underwent an NIH review by a panel of external experts. The proposal was not well received by the reviewers who did not think that there was anyone suitable at Boston University to do such a study, plus it would cost tens of millions of dollars. Reviewer 3 noted that "Genotyping 2000 individual DNA samples with 100,000 SNPs is likely to cost this group $300 million or more". Furthermore, the panel decided that "The budget estimates for the genetic analyses appear naïve and inaccurate. The statistical approach appears pedestrian but adequate. The expertise in molecular genetics does not appear adequate to accomplish the tasks proposed". There you have it – the reviewer dismissed my marching style as pedestrian – even the army never went that far! Not only was the approach not funded, but 6 million dollars was held back from the Framingham Heart Study contract on the reviewer's recommendations, perhaps to fund someone outside Boston University more qualified to perform the genetic study. Not a great way to please the boss!

A number of committees were then formed to control the operations of the genetics laboratory. One committee oversaw requests from non-Framingham investigators for DNA. They would then approve the use of the relevant subset of the available measured traits. The aim of these studies was to relate variations in DNA sequence to disease, thereby testing whether a particular gene played a role in the outcomes. If so, some gene variants might decrease the risk while others could increase it. We built standardized plates for distribution of different DNAs. The DNAs were extracted from immortalized cell lines prepared previously from blood donated by the participants. Only those DNAs where we had the consent of the participants were distributed. Whereas most people did not mind academic studies, there were a few who do not agree that commercial companies should have access to their DNA. All samples and data were deidentified, but in a way that we could collate findings once they were returned to the FHS. All DNAs were handled as mandated by the agreement signed by the recipients. We also developed protocols for amplifying any available DNA when cell lines were unavailable.

Another committee was set up to decide what DNA variants would be typed in the genetics lab, one at a time. I was all for doing as many variants as we could with the new technologies under development. Such an approach required careful statistical analysis because doing hundreds and thousands of tests would result in a lot of positive results from chance alone. Clearly, many of those positive results were wrong and would generate false leads. The analysis also required automating many of the steps involved, putting the statisticians and investigators in a role different from their existing ones. Not surprisingly, there was not any support from the FHS people for a genome-wide study, both because of the multiple testing problem and because of the scathing review received for my proposal. Phil was even less impressed when he realized how expensive the genetic tests cost to run.

Then the new Head of Genetics and Genomics at Boston University, Mike Christmas, also an MIT alumnus, arrived. He had trained in yeast genetics with Jerry Fink and was beginning to transfer some of the findings to study chromosomal changes in cancer. His department was located on the same floor as the genetics laboratory. I discussed with him the challenge of a genome-wide study using the FHS data. He was excited about the opportunity as he thought the work was timely and could become a signature program for the new Department and for Boston University as well. He also had the startup funds to initiate the study.

We started to plan the work. How would we overcome the multiple testing problem? We needed a robust method to solve this issue. The approach we selected was based on the work of Nan Laird at Harvard, along with a junior faculty member, Christoph Lange, who worked with her. Rick Myers had suggested that I chat with Nan, given her experience with mapping genetic variants in families. We adopted a two-step strategy. First, we applied general statistical tests to look for an association between trait values and offspring genotypes. Those genotypes were inferred from those of the parents. In the second step, we performed tests that Nan developed using the genotypes directly measured in the offspring, not those imputed as in the first step. The two methods were indeed statistically independent of each other, ensuring that the two-step approach was valid [51]. We then selected the top associations identified by the first step, knowing that many results were false positives and a product of multiple testing. We could then use the family structure to identify a genetic effect. If the trait was truly associated with a genotype, then gene variants should segregate with the trait values. Crucial in this endeavor was our ability to follow transmission of a genetic variant from parent to child.

By limiting the number of associations carried forward from the first step to the next step, we could manage the multiple testing issue. The limited number of tests performed reduced the chance of a false positive test. By running only 100 tests in the second step, rather than 100,000 as in the first step, we could easily identify the false positives. The remaining hits could be tested for replication in a different study group.

The last time I attempted to discuss this approach with Phil Wolf was in the corridor outside his office. I began, "Phil, do you have a moment to talk a little more about the whole….". He looked at me as I spoke, turned around, paused as he might say something, then walked into his office and closed the door behind him, never saying a word. He was not going to give permission for us to do this study. And we were not going to proceed unless we had permission.

Without the support of the FHS investigators, the only way to access FHS data was from the public release version that was two years old. The NHLBI had put this process in place as they had received numerous complaints about the lack of independent access of investigators outside the study to the FHS resources. This process involved one of the committees I described above and that I had helped formulate. We established the necessary protocol. Each step came with checks and balances to protect the confidentiality of participants and their families. I was able to help Mike Christman with his application. We applied and he obtained approval for the proposal. I am sure that it took more than just a show of hands to have the application pass through the Committee. I was not involved in any way in the decision, not part of the meeting, nor privy to the conversations that followed. I was, however, involved in what happened next.

During the subsequent fallout, I was fired from the Study via an email sent around 2 am. The email was not addressed to me, but I was copied on it. The email was sent to Mike and stated that I was transferred to Mike's Department. There was also a letter sent to Mike from Phil about the genotyping project that had just been approved (dated May 18, 2004)

"I know that a project like this has been a dream of Alan Herbert's for several years. He has been a consistent advocate of research along these lines. It is clear that his career as a scientist would be better served if he were in your department rather than Neurology. Therefore, in the spirit of cooperation that we hope to foster, I would agree to the transfer of Alan's primary appointment from Neurology to Genetics and Genomics" (Figure 5.2).

FIGURE 5.2 Mike Christman, then head of the Department of Genetics and Genomics at Boston University, and Philip Wolf, then Head of the Framingham Heart Study: at the interface between molecular genetics and epidemiology.

According to the email I received, I was to vacate my office immediately and that there would be people there to assist me in the morning. That day, Mike's crew appeared and helped me move from my office to a new one on the floor. Rick Myers, who had heard the news, offered to put me on his payroll, but that was not necessary.

I didn't really have time to react – I had to clear my personal files from my FHS computer and focus on the genome-wide study. In one sense. I felt that I had just escaped before the FHS doors could close on me, just as Phil's door had closed on me the last time I had tried to engage him in a conversation about the genome-wide study. My tenure with FHS was about to expire one way or another. As Susumu might note, the outcome was not an accident. This exit provided a clean break between the science and the politics, at least for the next part of the project. Full steam ahead as we had permission to proceed with the science. Phil had clearly put some serious thought into how best to advance my career as a scientist, probably with the advice and consent of the Boston University legal team.

There was one subsequent meeting in which the Neurology faculty met with the Genetics and Genomics faculty, nominally to discuss the science. FHS investigators and members of our team were present. I was hoping that we could put a framework in place where the contributions of all the parties were appropriately recognized. It did not go well. Mike presented our progress and nonchalantly asked what the FHS investigators would "bring to the table". A poor choice of words, as the contributions of the FHS investigators had made our study possible. It was like mixing oil with water. The epidemiological approach of owning the data versus the molecular approach of getting the results to press were not easy to align. The meeting ended soon after it started. No one had anything further to say in such a public forum.

There were subsequent meetings between Mike and Phil mediated by then-acting Boston University President Aram Chobanian. In the blue corner, you had Phil who was bringing in tens of millions of grant funding per year, and, in the red corner, you had Mike, with a plan. An agreement was reached whereby Genetics and Genomics personnel would share the genotyping data and be listed as co-authors on any subsequent FHS publications. The FHS investigators did not honor that arrangement. The medical school administration did not intervene. Indeed, the new dean Karen Antman was listed as an author of one of the FHS publications based on the genotype data we generated. However, the FHS investigators did kindly note in their first paper that they had replicated our initial findings and thanked us for providing them with the genotyping data.

Once we had the FHS data, we had to deal with issues on how the measurements from each examination were named and how they varied from one examination to another. The knowledge of all these details and the reliability of each measurement over time was why FHS felt they should guide any study using their resource. They were sure that we would fail without their help, so they were confident that we would need to involve them later on the terms that they set.

Of course, I had some insight into the problems with the exam data, having sat around the table playing the age-old game of bashing the p-value until the results were statistically significant. That outcome meant the p-value must be lower than the magical value of 0.05 (i.e., that the chance of concluding there was an effect when there was none was 1 in 20). Of course, there were many adjustments to made to the variables included in the statistical models until the result was finalized. Despite the way it sounds, the analyses performed by the FHS group were very carefully performed. The work was usually published in a highly ranked journal and so care was taken to ensure that the conclusions were robustly supported by the data. The lessons learned at FHS were applicable to many other communities and very influential in guiding public policy. Slow but steady progress.

Our challenge was to develop a way to organize the data collected over multiple years using techniques that changed from one exam to another. We had to perform our own quality control and structure the variables so that the traits we analyzed were based on large amounts of data rather than on small numbers. Marc Lenburg in the Department of Genetics and Genomics was critical In this endeavor. He helped design and build the computer cluster necessary to perform the analysis of the data. He also sourced the database where we could easily access the data we needed, something not possible for the FHS investigators, given their restricted access to variables outside their turf. Norman Gerry set up the genotyping, using Affymetrix chips where a hundred thousand DNA variants could be typed at once, rather than doing each one at a time. The approach was inspired by the miniaturization of transistors by the computer industry so that millions could be printed at a time on a single silicon wafer. I initially set up the statistical analysis in S+, for which Boston University had a license, then switched to the opensource R-project fork. The focus was on getting the job done. Sue Seigel, President of Affymetrix, saw the value of the work and helped reduce the overall cost with very favorable pricing for reagents. Christoph Lange supplied the family-based software that we used to perform the analyses run in parallel on the Departmental computer cluster. Starting from scratch with empty laboratory space, we performed the entire project with a budget for equipment and reagents of less than one million dollars.

The family-based approach was possible because of how FHS was designed. At the time we arranged the study, there were two generations who had participated. Now, there are three. We replicated our findings with the FHS data through collaboration with other studies. The paper was published in the journal *Science*. I was first author and Mike was last [52]. The paper has over 900 citations according to Google Scholar. None of the FHS investigators were listed, although their contributions were acknowledged. And no, the study did not cost $300 million as predicted by reviewer 3. In fact, I don't believe that it cost NHLBI a dime beyond what they had already spent, other than those expenses related to their administration and their peer review processes.

After we demonstrated the feasibility of performing large-scale genotyping to discover genes involved in heart disease, FHS started billing itself as a genetic study rather than an epidemiological one. The BU [Boston University] Today headline of February 9, 2006 read "Framingham Heart Study leaps into genetics". The article featured Phil "at the cutting edge" of this new science. The FHS data were also now accessible directly to outside investigators for analysis and to use in their own research.

Although we were able to map genetic variation across the entire genome, we did not find any of the very many measured traits which were affected by ADAR1. This approach did not lead to any new insights on the biology of Z-DNA. No luck there. Interestingly, there is now evidence of involvement of ADAR1 in atherosclerosis and cardiomyopathy. It was not possible to make these associations statistically in the FHS given the incidence of disease and the methods we used.

The politics of our success was ferocious, as if that is not already evident. There were the personality clashes involved and the angst of the old regime feeling that their turf was being invaded. The primary NHLBI investigator in the study, Dan Levy, said that we "kicked their ass", despite the expectation that we would fail. Out of the subsequent turmoil, a decision was made to perform any further genotyping under the auspices of the NIH in Bethesda rather than at Boston University. We had originally planned to genotype the entire cohort as part of a consortium, with funding from industry and with data accessible to all while ensuring that the wishes and privacy of the participants were respected. Dan Levy magnified the concerns about privacy issues arising if there were industry funding and was a major influencer in moving this work to the NIH. This outcome also involved new funding for the NHLBI investigators, such as Dan and Chris, to perform their own studies. Dan was given his own budget, free at last to undertake studies in parallel with those conducted by the Boston University FHS investigators.

Another group of people was also very unhappy with us. They were from The Broad Institute which had contributed a significant amount of data to the sequencing of the human genome and to generating maps of genetic variation across the chromosomes. They had managed many steps in the rollout of genome-wide association studies (GWAS) that were designed to map the genetic variation to a range of different human traits and diseases. We beat them to press.

It was likely, though I don't know for sure, that someone from their team reviewed my initial proposal for the FHS genome-wide study and authored the scathing review. They clearly thought that they had taken care of business. Consequently, the Broad team was completely blindsided when they learned that we actually had done the study. That was not possible, they had thought. David Altshuler and Mark Daley, then heading the Broad GWAS group, first heard of our progress at a Gordon Conference, a meeting where a small group of select academics meet to discuss work in progress. I had not been invited to speak at the meeting but, due to a cancellation, I was able to give the last talk in a session devoted to the Broad roadmap for GWAS. When I presented our data along with replication from other studies, David Altshuler was very generous in his praise of our results. We had done everything correctly from the design to the replication. The result only confirmed their overall strategy. I am sure David must have had some inkling of what was going on as he was best man at the wedding of one of our collaborators, Joel Hirschhorn from Boston Children's Hospital. So maybe his response and his remarks were already prepared as he already knew of our success. Joel was helping us in the replication studies and continued the work by forming the GIANT consortium to look at genetic determinants of human height in over 5 million individuals.

Meanwhile, we were in no man's land, targeted by all. Although we made our data freely available, you will be hard pressed to find any mention of our contribution to the dbGAP database where it is now stored. During the entire conduct of our work, we followed proper procedures and complied fully with any and all ethical guidelines put in place by NHLBI and Boston University. Our aim was to advance knowledge. However, we were, as they say, off-message. The acrimony within Boston University was not helpful to fulfilling the NHLBI mission. Boston University was quite simply described to me in a private conversation as "dysfunctional" by David Altshuler, who told me that it was not Boston University's place to do the genetic studies in the first place. They were there to collect the data for others to analyze. The fallout only strengthened his opinion. The FHS team just kept stoking that fire. One colleague of mine in Bioinformatics at Boston University was warned that if he collaborated with me, then the Framingham Investigators would not collaborate with him. I am sure Christoph Lange received the same message. Mike Christman left Boston University soon after. I should have done so sooner than I did.

The politics also worked against Phil Wolf. Funding for the FHS was cut by 40% in 2013 and he retired. Not the way I imagine he wanted to leave his club. Doubtless, the review involved some of the same committee of rivals who refused to support the GWAS proposed a few years earlier. In the end, while defending his temple against us, Phil had his pocket picked (his term, not mine). While distracted, he was, using his same parlance, an easy mark. It was a sad outcome for all who had made Framingham the success it was. Overlooked was Phil's rescue of the study during an NIH funding gap. The 2013 cuts mostly affected the people who had worked tirelessly with the Framingham Community over the years. Together with the participants, they had built a scientific study embraced and empowered by the proud families who literally gave heart to the Study. The study now continues under different leadership. It has recently received funding to study Alzheimer's disease.

The lesson for the administrators involved at Boston University was that viewing science as a way to maintain funding will always fail. Prior to my association with the Study, Boston University managers had failed previously in their attempt to spin out a private for-profit company called Framingham Genomics. The deal had the blessing of the-then NHLBI Director, Claude Lenfant, but was quickly canceled. As described in the article by Naomi Aoki in the Boston Globe, some participants "expressed objections to a company profiting from their voluntary participation in the study" (Boston Globe December 29, 2000, p. C5).

Good administrators know how to facilitate the science rather than treating research as a profit center. They raise funds to build the future of their institution. They enable their faculty. Their mission is to hope that the successes they help facilitate exceed their wildest forecasts. Good administrators engage alumni as donors to build support for the institution and in doing so add value to the degrees awarded to their graduates. Good administrators motivate donors through a vision that advances our capabilities as a society. They proudly graduate their students without debt so they can focus on making a difference. With a strong faculty and strong alumni, good administrators build a strong community.

Good administrators also fund their own salaries. Their goal is not to hire sufficient faculty to ensure that there is sufficient grant overhead to fully fund the management's lifestyle. It may sound radical but shouldn't administrators fund-raise to pay their own salary, say in the form of a finder's fee? At the same time, should they not collect funds to pay their faculty so that grant money can be used for research? How about the administrators providing financial support for their students rather than building dormitories that could double and be priced as luxury hotels?

THE MERCK EXPERIENCE

I was recruited in 2013 to Merck & Co., Inc. in the newly formed Genetics and Pharmacogenomics Department, headed by Robert Plenge, recently recruited from Harvard Medical School and the Broad Institute. The mission was to use the published genome studies of the type we performed in FHS and any other genetic information to identify proteins that were causally associated with disease. The process was called target identification. The chosen targets then would be carefully annotated and assessed for their "drug-ability". Then a "hit" finding program would start to find the drugs that would change patient outcomes.

There was plenty of data to analyze and plenty of opportunity to build strong collaborations between departments, each with their own skillsets. Unfortunately, the Department was treated by the more established groups as the latest fad. We were given resources unavailable to the others. The existing crew was quick to note that some of the initial targets identified by the new folks were very hard to treat with drugs, using conventional approaches. They had the experience to know a good target when they saw one. Developing a "hit" into a "lead" compound took a lot of resources, so why go after something that was not exploiting all they knew from previous screens, all the insights they had from previous campaigns, and the extensive library of proprietary compounds previously developed? In short, why not stay with the known, rather than veer off into the unknown?

Science at a drug company is now a rather unique profession, where it is curious that curiosity is a negative attribute. Certainty is certainly the certain measure of success. Any deviation from a well-beaten path could result in a question from the Food and Drug Administration that would delay a product launch and subtract billions from the bottom line. Even worse, such a delay could tank the stock price. These enterprises are serious business. Risk reduction is the key to survival for any large company. In the battle, it is extremely strategic for a group leader to mirror the competition. If a project fails for the inside team, it will also fail for the competing companies as well – so, even with failure, the local players are safe as they clearly show that they performed as well as others in the industry. Everyone is happy. Failure can also increase your bonus because no money was wasted on developing a program that everyone now knows wouldn't pay off. The reverse is also true. If the competition succeeds, your team will also succeed. People are even happier as the success shows that we are as good as those other guys. Of course, a company is a business and successes are measured in terms of dollars earned rather than advances made. Everyone must respect the bottom line. At that time, Merck & Co.'s top money-earning drugs were based on a solid understanding of the biology, but the actual proteins targeted were not genetically validated.

When I joined, everyone at Merck & Co. was waiting for the next blockbuster. Merck & Co. was founded originally in Germany in 1668. Its first blockbuster was morphine, launched in 1827, the oxycontin of that age. The American Merck was stolen from the German Merck during World War I, or rather was expropriated in 1917. When I joined, the bets were placed on checkpoint inhibitors for the treatment of cancer. It was hoped that the one under development, the anti-PD1 drug launched as Keytruda, would deliver. The pipeline of new drugs had recently clogged with nothing but a drip coming out. Anti-PD1 turned out to be a great success story. It showcased Merck & Co. at its best and is described by David Shaywitz in the July 26, 2017 *Forbes* article entitled "The Startling History Behind Merck's New Cancer Blockbuster".

The blockbuster asset was originally developed by a small Scottish biotech called Organon. They had rolled the dice and initiated a number of new initiatives, but they were purchased by Schering-Plough for the following Organon products: FOLLISTIM/PUREGON, a fertility treatment; ZEMURON/ESMERON, a muscle relaxant; and NUVARING and IMPLANON for contraception. Judged as lacking potential, the anti-PD-1 antibody-producing cells were frozen away. After Schering-Plough was purchased by Merck & Co., the PD-1 program was again judged as lacking any potential for blockbuster status. The antibody was then listed as an asset for sale.

Timing is everything. Just then, before any buyers emerged for the Merck & Co. antibody, Bristol Myers Squibb (BMS) announced success in an early-stage clinical trial with their PD-1 antibody. The antibody amplified immune responses against malignant cells. The mechanism of action was novel. The Merck & Co. program was immediately resurrected. Why not? Those people at BMS had shown that the Organon antibody was now an asset. There was proof of clinical efficacy and a potential blockbuster.

The antibody-producing genes from Organon were given new life in a new body. They were transferred to a cell engineered to produce antibodies at levels equal to or better than the industry standard. Merck & Co., who had previously sourced their antibodies from their partners, then co-developed a whole new production method using disposable bags that could be used at multiple sites under the same manufacturing license as the process employed was the same at each facility. Merck & Co. moved quickly to launch their own clinical trial. They were behind at the start, but ahead at the finish. Their Phase 1 trial would eventually involve 1,235 patients and lead to many FDA approvals for different cancer indications [53]. No one previously had ever had a Phase 1 trial that large. Remember, a Phase 1 trial is only intended to show a product is safe for use in humans and to establish dosing for the subsequent Phase 2 and Phase 3 trials. Merck did start small. But each result led to another question. Was it safe for this indication if we went higher in dosing? Was it better if we dosed only every two weeks or every four weeks? Was it better if we used antibody shipped in solution rather than freeze dried? Was the dose that worked in this cancer the same dose that would work in more difficult-to-treat cancers? Was this measure of clinical response accurate? As the numbers accumulated, it soon became apparent that the treatment was effective where others had failed. Was the FDA going to block the path to the clinic? Absolutely not. Although Merck & Co. was not the first to test antibodies against cancer in humans, they were first to launch the commercial product. The coordination of different assets, especially at hospitals conducting trials, was quite phenomenal.

BMS, of course, cried foul as they had pursued a more conventional drug development program that progressed from Phase 1 to Phase 2 to Phase 3 to clinical registration. Of course, BMS did have a rather broad patent on this approach. Was that a problem? BMS answered the question rather quickly by suing Merck & Co. What was Merck & Co to do? Easy, just sign a licensing agreement to pay for use of the patent. Merck paid $650 million up front and a royalty stream itself pegged at 6.5% of Keytruda sales through 2023 and 2.5% through 2026 (Tracy Staton, January 23, 2017, Fierce Pharma).

But why would BMS so easily give up its monopoly? Was it the bad public relations that a lawsuit would produce? What would you think if you, a relative, or a friend had cancer, and access to a life-saving therapeutic was restricted by lawyers, arguing over who had the right to sell it? There was also a risk that at some point the patent might be invalidated. When all added up, it was better to settle for money and protect the patent. The agreement would also maintain the barrier to other drug companies entering the fray, even if the patent was not valid. So, after one press announcement noting the intent to litigate, the issue quickly settled. There was payment of the licensing fee and a royalty stream based on future sales. Each side had too much to lose by going to court. This outcome helped protect both Merck & Co. and BMS as they expended money to develop their programs further. It was also good for the stock of both companies as the success of one in a clinical trial using their anti-PD-1 antibody increased the market value of the other. Just good business. Of course, the top C-Suite executives at Merck & Co., like Ken Frazier, the Chairman of Merck & Co. and Roger Perlmutter, the President of Merck Research Laboratories, received huge bonuses for being so clever. Best of all, the patients benefited from the rapid deployment of these powerful new drugs for the treatment of cancer.

Merck & Co. executed their strategy perfectly. If it worked once, why not try it again? First, identify an asset with external validation, then acquire it or find an equivalent asset. Such acquisitions were a frequent event during the time I was at Merck & Co. As Tony Siu, a colleague from Merck & Co. noted, it was interesting that the announcement of these new bold initiatives to expand the drug pipeline often came just before a press release describing the failure in the clinic of a previous bold initiative. The bounce in stock price from the new deal offset any losses incurred from the bad news. Why not? The compensation of management was based on stock price. Just like stockholders, the management only like shares that increase in value.

Internally developed molecules also had their success, although sometimes an external asset was acquired or licensed to bolster that particular program. Once in development, the necessary steps were well defined. Each milestone involved a go/no-go decision. Any discussion focused on getting to go was championed. Anything else was considered a distraction. These programs were based on a matrix design.

Individuals from different departments sat around the same table tasked with making it all happen. These were the horizontal rungs of the matrix. The seven vertical layers consisted of management. That was the degree of separation between junior scientists and Roger Perlmutter. Quite a layer cake. Everyone had deliverables to hand to their manager. To help get that message across, there were courses so that you would know how to be a good corporate citizen. Implementing industry standard practices that were approved further up the line was mandatory. There were also lessons on the use of the company trademarks and the protection of intellectual property. At no stage were you to confuse Merck & Co. or, as it is known in other places, Merck, Sharp and Dohme and sometimes MSD, with the original German Merck. Also, reviewing information on harassment policies and acceptable behavior was mandatory with frequent refresher meetings scheduled.

Each Department had their legacies that were part company lore and part industry standard. For example, much of the bioinformatic analysis at Merck Research Laboratories was performed using MatLab, so I learned to code with that even though my previous experience was based on the open-source R statistical language. Then, we also had communications handed down from higher up. Often, the word would pass down the line that a particular project had been canceled. The reasoning was not always clear – whether it was the science, competitive realities, or some other reason based on the business case. A lot of time was spent around tables trying to divine projects that might gain support, rather than doing anything experimentally to develop an idea further. There was more focus on opportunity cost rather than opportunity lost. Surviving the matrix was an everyday challenge. No one at any level wanted to be the author of a project that might fail for one reason or another. Those who killed projects early were also rewarded as this allowed for better use of resources. The matrix had so many aspects to it that many processes were just referred to by a three-letter code. It was fun trying to guess what the letters stood for. There was over a thousand of these codes. It was best to go to the website that translated this code into words and to bookmark the website as finding it was not an easy task, there were just too many variations to remember.

If you did fit into the matrix, then you would be separated from it, as in "You are being separated. It was nice working with you". There were frequent reorganizations that involved people coming and going. The managers would huddle and decide, then go through a check list provided by Human Resources that was designed to prevent any future legal claims by a separated individual against Merck & Co. On that list were responses to any of the issues you had discussed confidentially with Human Resources. There was no union guy on the premises to help mediate any problems. Separations in Genetics and Pharmacogenomics were implemented three times in the three years I was there. You served at the pleasure of people a few layers up the matrix. This meant that it was very squishy at the bottom of the layer cake. No one could ever be sure that they were standing on solid ground.

For those separated, there was a payout and paperwork to ensure that everyone understood the *quid pro quo*. So, was I the right fit for the Merck & Co structure? No, not at all. That didn't stop me from trying to identify promising targets for drugs. The oncology group with whom I initially worked had access to a large amount of RNA expression data from tumors collected at the Moffat Cancer Center. The profiles had been repeatedly analyzed using the standard tools to generate gene sets that were correlated with outcomes. Of course, the contents of each set depended on the cut-off value selected for the correlation. The process was also complicated by the noisy nature of the measurement, with variations arising from when, where, and by whom the data collection and cleaning was performed. When at Boston University, I had played around with a similar problem. The question is easily stated: "How do you extract the signal buried in so much noise?". At Boston University, I was curious how a zebrafish was able to recognize and track food three days after fertilization. Clearly, there were too many neural connections to specify genetically. Believe it or not, I thought it was related to the more general problem of whether statistics can really help in the analysis of genetic traits. I found an algorithm that made it possible to identify key attributes of an object that were above a noise threshold. The procedure relied on repeatedly transforming the data, using a set of randomly generated masks to view the underlying data structure from many different angles. The intuition is that objects that are truly connected remain close together in the projections regardless of the mask applied. In contrast, the signal from noise becomes dispersed throughout the different spaces created. It is an idea derived from the field known as compressive sensing that is widely used to eliminate noise from images. Essentially, the approach allows the reconstruction of a perfect image from many degraded copies of that image. In my case, I fed in random noise and looked for output that deviated from a random output, not so different from the statistical analysis used for code breaking by cryptanalysts. In my case, I worked with a standard set of numbers from the MNIST (Modified National Institute of Standards and Technology) database). It was possible to obtain over 99% accuracy using just 30 parameters, clearly within the range of genetic encoding, and far less than the many thousands used in the various neural networks in use at the time.

At Merck & Co., I tried this approach on the cancer datasets that measured RNA levels for thousands of genes in thousands of tumors, hoping that the method would allow me to find molecules that were strongly associated with PD-1 and could be drugged in a way that made anti-PD1 antibodies work better. I turned up four such molecules, namely IL4I1, IL15RA, IL32 and FOXP3. IL32 induces FOXP3 that has been associated with T cell that suppress the anti-tumor immune response while IL15RA plays a role in activating antitumor T cells. The IL4I1 finding was novel and was replicated in other data. Published studies had also established an immunosuppressive role for IL4I1, but the target was not an any Pharma's radar screen. You might have thought this was a good thing. However, there were a number of strikes against. First, at the time, there was no supporting genetics (remember, I was in the Department of Genetics and Pharmacogenomics), so it was not a score for my team. Second, by working with a different department to put a program together, I was competing for downstream resources that Genetics and Pharmacogenomics might want to claim. Third, no one understood the algorithm – no surprise there because it was not standard Pharma fare. My results were subsequently validated by other members of my team using the standard tools. It is usually easier to check that an answer is correct rather than to solve the original problem. Those team members with access to the clinical data showed that the expression of IL4I1 was a strong predictor of clinical response to anti-PD-1. My reward was to receive a training memo designed for those who were outside the straight and narrow lines drawn by those above. The mission was to enhance the Department's standing (and hence its leadership team) rather than on discoveries like the one I made. Incidentally, my immediate manager also received a training memo, likely because my transgressions happened under his watch.

I was transferred to a new manager. The task of my new manager seemed to be to document how I spent my day and to assign tasks that were rather menial. He thought that it was important to extract data from tables in PDF files of published GWAS studies. Of course, that was not defined as harassment under Merck & Co. guidelines but rather it was described as a mission-critical initiative. At one point, I was instructed by that manager's manager not to use my computer for anything else other than the tasks assigned to me by the manager, even if I used the computer out of hours. The supervisor, Caroline Fox, I had known before Merck & Co. She was one of the p-value bashers from the embedded NHLBI team at Boston University, earning a Master's Degree in the process before publishing GWAS studies based on the genotyping data generated by the Department of Genetics and Genomics when we performed our FHS study. The messaging was hard to miss. There was no point being upset. My game plan was just the standard one everyone else in my position followed. I just waited for the next reorganization as I knew I would be handed my boots plus a check as part of the separation. If I had left before then, it would have been without a check. Others who decided to leave the company for one reason or another would also time their exit, usually to another company, to ensure that they received the annual bonus on the way out the door. The strangest feeling that I had when I walked out the door for the last time late in 2016 was that any scientific achievement made while at Merck & Co. was now locked within the doors behind me. There was nothing to take forward.

A few months later, early in 2017, Robert Plenge's departure from Merck & Co. was announced. I guess it did not work out for him either. My remedial manager also hit the departure lounge soon after. Caroline Fox is still going strong. She must be doing her job well as a corporate citizen, producing pharma-perfect p-values to please all! My first manager at Merck & Co. was able to use the IL4I1 program to show the value of his contributions. Merck, Sharp and Dohme was issued patent WO2022/227015A1 for over 278 potential IL4I1 inhibitors in November of 2022.

I was interviewed for jobs at major Pharma afterwards, but it was clear to me that "rinse and repeat" would not be a great idea. In the end, it came down to two questions I would ask during the interview after all the *pro forma* exchanges. My first question to the interviewer was how long had they been with their wonderful company. Usually, the response was somewhere between ten and twenty years. That meant that that person had survived a number of reorganizations and obviously had an ability to catch the right wave. The next question was to determine which of their talents allowed them to survive so many upheavals that lesser mortals were less able to endure. I then asked "You must have worked on drugs that made it to the clinic? How many of your programs have advanced that far?". The usual answer was none. I then waited in silence to see what would be said next. Commonly, the next statement was that their job was one of support as they had no direct involvement in drug development. They were team players, with no role in the outcome. Another response was that, in the big Pharma business, you needed a high tolerance for failure. After three such experiences with other large Pharma, I decided managed failure was not for me.

Not a problem! After Merck & Co., I was financially able to do my own thing. In putting various job talks together, I also searched for new approaches to cancer therapy. I did not have the tools at my finger tips that I had at Merck & Co. There was, however, a treasure trove of data out there from a number of NIH-funded mega-projects, all technology driven. Those are the projects the NIH likes as the outputs are easy to quantitate in terms of "bang for your buck". They also ensure a modicum of quality control, although that analysis is limited by the algorithms available at the time. The projects also keep the workforce employed, and provide training for a future role in industry. Although sold as a discovery resource, their value for me is that they provide great way to test the validity of a particular hypothesis. I now had the skills to analyze them. I could ask many different questions designed to give a clear yes or no answer. If the answer was no, I could move on to a different explanation of the data. To help perform this analysis, I also had an amazing amount of open-source software available. So, while I could say that I was starting again with nothing, that was not true. The challenge was to find value in what was already freely available.

Once again, this new phase of my life started with bare shelves. Soon, I had the basic infrastructure in place: a computer loaded with open-source software and a fast internet collection. Then, I started looking more generally for immunomodulators that affected cancer outcomes. I was focused on approaches for targeting those tumors that hid in plain sight. That meant that they did not elicit an immune response against them, even if they made mutant proteins that the body should recognize as foreign. I was going after cold tumors. Where to start?

Why not check the literature to see what others had found? I came across a paper from the Afshar-Kharghan laboratory showing that deletion of a gene called C3 (complement component 3) prevented tumor growth in some cancers [54]. What is C3? The gene encodes a protein that arose as part of an ancient immune system, perhaps the first defense multicellular organisms had against invasive pathogens. When activated, complement C3 bonds irreversibly to proteins and carbohydrates on biological surfaces. It sticks like superglue. What happens next determines whether or not an immune response occurs. With pathogens, the C3 protein undergoes breakdown into a form that turns the body's defenses against the pathogen. Host cells protect themselves from attack by turning C3 into something that inhibits the immune response. They have a whole set of proteins that prevent an attack against self. We know because variants of these proteins that do not function properly are associated with the development of autoimmunity. Tumor cells exploit C3 to prevent immune responses against them. That was the implication of Afshar-Kharghan's finding. But how?

Tumors coat themselves with the form of C3 that prevents an immune attack against the abnormal proteins that they produce. By doing so, cancer cells grow unchecked by the immune system. I thought that this possibility was interesting enough to test experimentally. Could we coat them with the inflammatory C3 product, called C3d, that stimulates the immune system? The easiest way to do this was to make a DNA encoding for C3d and inject the DNA into tumors. The tumors do the work of making C3d and placing it on their surface. All I need to do was add something to the C3d protein to take it to the tumor cell surface. That was easy as that involved adding a small sequence to which the tumor would add lipids. These fatty tails would direct the C3d to the cell membrane. A back-of-the-envelope calculation showed that this experiment could be done for less than $15,000. So, it seemed worth a shot. Initially, the plan with the first experiments was just to check that the tumor was expressing the C3d. We would take the tumor cells injected with DNA from the mouse and check how much C3d was on the tumor cell surface using a fluorescence-activated cell sorter (FACS) for the analysis.

A few days after the first DNA injection into tumors, I received an email from a scientist at the Contract Research Organization in China. It was Monday morning (October 8, 2018). I had not had my first shot of caffeine for the day. The email had lots of yellow highlights. The heading was "Experiment did not work". My immediate thought was that there were problems with the protocol. Money down the drain, though it was definitely worth trying. I closed the email without reading further and had my coffee. Now more alert, I examined the body of Zegen's text more closely. He wrote: "We found the tumor of mice in group 2 start to shrink after treatment. I afraid there are not enough tumor cell for FACS analysis tomorrow even if pool three tumor to one sample." The treatment had caused the tumor to shrink! The result was spectacular and unexpected. I was really excited. Could it be that all the data analysis and hypothesis testing *in silico* had actually delivered a worthwhile result? Did I have a new therapeutic?

Of course, my second reaction was that the result was a fluke and due to a technical error or some other artifact. Further experiments revealed that the effect due to C3d was real, but was not optimal. Working in DNA makes it easy to make variants in the protein sequence to probe the mechanism of action. It was easy to test what features of C3d were critical as there are a lot of genetic variants of C3d that cause autoimmunity, a disease where the runaway immune response destroys self-tissue.

With the critical features established and efficacy in a preclinical model (Figure 5.3), I filed patents citing the results. Naïvely, I thought having a novel mechanism of action, proof of principle, and a patent would enable further funding for the work. It didn't. Not from venture capitalists and certainly not from the NIH. Those in the betting parlor either wanted a sure bet or a company run by a team of proven winners who had done this type of product development before. Furthermore, the approach was not on the big Pharma shopping list. It was too early, meaning that the competition had not validated the strategy. Since the company I formed was a for-profit, mostly to offset the expenses against my personal taxes, not-for-profit funding sources, such as disease-focused charities were not interested either. In addition, I was not able to fully disclose our technology because of patent issues. The problem is mostly with the European Patent Office. They have a very liberal definition of what constitutes a public disclosure. Getting academics involved was difficult given

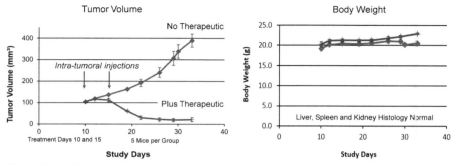

Clear cell Renal Carcinoma 786-0 Xenograft with injection of therapeutic into established tumors > 100 mm³

FIGURE 5.3 Tumor rejection with the C3d therapeutic I developed that was encoded in DNA and delivered by injection into an established tumor. The red line shows treated and the blue line untreated animals in a pre-clinical 786-0 xenograft clear cell renal carcinoma model.

the various intellectual property issues involved with their institution. Many technology licensing officers only see the world with dollar signs in their eyes. Most universities were already working with their own startups or had other conflicts. Even after I published a manuscript disclosing the principle, but not the details, behind the discovery, there was still no interest. The paper has currently three citations.

Millipore-Sigma offered to sponsor lab space at LabCentral, a startup hub in Boston, but still there was no working capital. Multiple applications for funding by the Small Business Innovation Research (known as SBIR) were also panned for a number of unrelated reasons. I really don't know what it takes to obtain NIH funding. Certainly novelty, proof of principle, and pre-clinical validation are not sufficient. The science was well supported by human genetics and by the data generated. I published the mechanism based on our experiments in the *Journal of Immunotherapy for Cancer* [55]. Yet the scores by the peer-reviewers for each submission were all over the place, collectively out of the funding range. Without a doubt, it is another example where the army is out of step.

It seems that many academics have a similar problem. With the NIH funding rate so low, it seems that such academics spend 95% of their time writing grants and the other 5% on review committees ensuring that their field is funded. What else is their incentive for spending so many days a year reviewing? It is uncertain in my mind whether peer review is actually a better system than a lottery in which all qualified applicants are given an equal chance of being funded as bias is no longer a deciding factor [56]. The lottery addresses the concern with the current system, where the allocation of research monies favors low-risk projects and therefore established players who are good politicians but not necessarily the best at stepping out of line.

6 How Do You Know It's You? The Answer Lies in Your Z-RNA

"As a historian, I would say that these new explanatory models are highly different and much more complex than the simplistic models that were immediately proposed after the discovery of Z-DNA. I ought to change the subtitle of my article: Nature was opportunistic, but as always in a more intelligent and sophisticated way than what was initially imagined". Michel Morange, March 2, 2021 (email to AH).

"I considered my error: it was not to say that the first models that were proposed for the functions of Z-DNA were wrong. It was to deduce that Z-DNA had no functions, and that, in this case, Nature had not been opportunistic. Whereas it was simply that Nature had been very smart, and for this reason it required a lot of time to unravel the way it works". Michel Morange, May 28, 2021 (email to AH).

I had progressed from a medical education to a PhD in cellular immunology to a molecular biologist who mapped the first Z-DNA binding domain to a statistical geneticist who helped pioneer a method for mapping human genetic variation to phenotypes to a computational biologist identifying new therapeutic targets at a major pharma company. Despite having had one computer class at medical school where I had to write a Fortran program on punch cards to print out my name, I was now able to explore a vast expanse of data to discover Nature's secrets. At each stage of my career, I learned and developed methods needed to explore the art of using counterfactuals to disprove an otherwise sound and clearly stated hypothesis. I borrowed what I needed from other fields to address the questions at hand, not necessarily what other people in and around me were familiar with. I founded a company called InsideOutBio as a play on this approach to thinking outside the box. The name was also a nod to the Disney film of the same name that captures the endless eddies of emotions experienced as life praters on. Then, there was a series of fortunate events which led me back to the Z-DNA conundrum to help in the rapid unravelling of its biology. Quite an adventure!

After Merck & Co., I was a company of one with no duties to report, nor any obligations to an institution. I had time to spare as I waited for my contract research organization to run experiments for InsideOutBio on the complement therapeutic. Much as one might enquire about old friends or acquaintances to see how life worked out for them, I revisited my work on Z-DNA. Maybe something good had transpired or maybe there was something new there that I could use to light up cold tumors. Searching the literature, I came across a review by Michael Jantsch, who I knew from RNA editing days. I scanned the paper and saw the title "What is the

DOI: 10.1201/9781003463535-7

Biological Role of the Z-DNA Binding Domains in ADAR1?". The question surprised me. Our initial idea was that it was there to localize ADAR1 to editing substrates. So, was that not the case? Then, I began looking deeper into the happenings
since I was last active in the field. Sung Chul Ha, working with Kyeong Kyu Kim,
solved the structure of the B-Z junction using Zα to stabilize Z-DNA in one half of a
DNA duplex, while the other half of the duplex remained in the B-DNA conformation [57]. Only a single base was extruded at the junction between B- and Z-DNA. Zα
was crystalized, bound to RNA, by Diana Placido [58]. The result allowed Z-RNA
to be visualized for the first time and provided additional confirmation that Zα was
structure-specific. The only difference with Z-DNA is that initiating the flip was a
lot harder. More energy is required to disrupt the A-RNA structure due to the extra
hydrogen bonds formed by the hydroxy group on the ribose sugar of RNA but not
DNA (giving it the name of deoxyribonucleic acid). It was the reason why Z-RNA
did not compete with Z-DNA in my band shift assays. The Zα domain from ADAR1
could substitute for the Zα domain of vaccinia virus E3 protein, the Zα family member we identified in our first paper, to maintain the virulence of the virus [59]. The
Zα domain attached to a transcriptional activation domain could also activate gene
transcription [60]. All of these were interesting findings and not unexpected, given
all the work I and others had done previously. However, the studies did not provide
much insight into what left-handed nucleic acids did in a cell.

It seemed that no one was actively pursuing the biological role of Z-DNA further. Not a surprising fact as experimental science was not Alex's forte. One post-
doc, Alekos Athanasiadis, had continued on in his own lab to solve structures of Zα
family members encoded by different viral genomes [61]. Yang Kim had returned
to Korea where a Korean National Laboratory of Z-DNA had been established by
the Korean Government, presumably as part of a national campaign to make the
country more competitive in the biotech industry. Loren Williams and another post-
doc, Martin Egli, had crystallography labs that provided insight into the way small
molecules and water molecules stabilized the Z-DNA structure. They also branched
out into other arenas.

The reasons for the lack of progress in understanding Z-DNA biology were well
laid out in a 2007 article by the historian Michel Morange, who stated, "Z-DNA is
an example of a discovery made by accident, where, however, belief in serendipity
has so far led those who adopted it to a dead end". [31]. The article called out Alex in
particular and was written as an epitaph for the field. By 2007, the fruit had withered
on the vine (Figure 6.1).

As I pondered the situation, I wondered whether Michel Morange was making the
same mistake as others had in the past. He clearly equated an absence of proof with
a proof of absence. He knew of the "Z-DNA-binding nuclear-RNA editing enzyme"
but did not let that affect his conclusion. I was more than a little surprised by the
state of affairs.

I started digging deeper. I focused on any mention of ADAR1 p150 (a refences
to its 150,000 molecular weight), the longer form of the ADAR1 editing enzyme
that I purified and that included the Zα domain. I knew that this form of ADAR1
was induced by interferon, part of the response to viral infection, as published by

1983 RESEARCH NEWS
Z-DNA moves toward "real biology"
By G Kolata

 1985 RESEARCH NEWS
Z-DNA: still searching for a function
By Jl. Marx
Six years after the discovery of Z-DNA questions remain about whether it exists naturally and what its functions might be

 1990 **Are many Z-DNA binding proteins actually phospholipid-binding proteins?**
PRITI KRISHNA, BRIAN P. KENNEDY, DAVID M. WAISMAN, J. H. VAN DE SANDE, AND JAMES D. MCGHEE*

Journal of Biosciences **2007** **What history tells us**
IX.
Z-DNA: when nature is not opportunistic

MICHEL MORANGE

FIGURE 6.1 The waning of the Z-DNA literature. Was the quest for biological significance a lost cause?

Chuck Samuel's laboratory. That was one clue. Mice without the ADAR1 protein died at day 12 after fertilization as shown by Qingde Wang, Kazuko Nishikura, and Peter Seeburg. Jochen Hartner showed that interferon activation occurred prior to embryonic death, even though no pathogenic viruses were present [62–64]. That was another clue. Simone Ward and Chuck Samuel showed that it was p150 that was necessary for embryonic survival [65]. Mice that only made p110, the short form of ADAR1, also exhibited embryonic death. Another clue. Carl Walkey produced a mouse that had a form of ADAR1 that was incapable of editing RNA because of an altered amino acid change in its enzyme domain. The protein was unable to convert adenosine to inosine. The embryo still died, maybe a day later. However, if Carl bred these mice to another mouse strain that lacked a protein involved in the interferon response to double-stranded RNA (dsRNA), then the embryo survived [66]. Another clue. The dsRNA-sensing protein was called MDA5 (short for melanoma differentiation-associated protein 5). The protein filaments formed by MDA5 on dsRNA act as scaffolds to assemble other proteins that initiate activation of the interferon response. But there was nothing related to Z-RNA.

Putting these findings together showed three important things. First, that double-stranded editing by ADAR1 actually was not necessary for the normal development of mice: the editing-dead enzyme did not result in any birth defects, provided that the interferon activation was blocked by deletion of the dsRNA sensor MDA5. Second, when MDA5 was present, the editing of dsRNAs produced in the absence of a viral infection was essential to prevent embryonic death. Third, the Z-RNA-binding region of ADAR1 was also involved in this process.

Science does not always proceed smoothly and these clues were missed. As in any good mystery story, it is often only in hindsight that you see how easily the facts connect. To explain these ADAR1 findings, a mechanism is required to prevent normal cellular dsRNAs from activating an interferon response against self while allowing the necessary response to protect against virally encoded RNAs. A number of proposals were made over the years for how such a system works. Initially, it was thought that the cell does not make dsRNAs long enough to trigger an interferon response. The self-made dsRNAs were too short to seed formation of the extended MDA5 filaments needed to induce interferon production. Those types of long, dangerous dsRNAs had been eliminated by natural selection during evolution. In this model, it was thought that MDA5 could untwist any of the remaining short dsRNAs so that the filament would never form in a normal cell. Only the longer dsRNAs made by viruses would lead to the assembly of filaments that activated the anti-viral response (Figure 6.2). Clearly, the mouse experiments ruled this out possibly as dsRNA long enough to activate the interferon response certainly existed in non-virally infected cells. Another proposal suggested that the host RNAs were modified differently from the viral RNAs. It turns out that viruses are very adept at repurposing host enzymes to make the changes necessary to protect the viral RNAs. The viruses can then modify their RNAs in just the same way as the host does, allowing their replication to proceed. If protection by inosine formation was an important mechanism to protect against interferon responses, then viruses would also adopt this strategy. So, it was unlikely that differential modification was the mechanism at work. Another idea was that adenosine-to-inosine conversion of host transcripts prevents MDA5 filament formation by destabilizing dsRNA. Contrary to this model, editing can also increase the stability of dsRNA. The editing of an adenine mismatched with a cytosine on the other strand to produce inosine creates a base pair that stabilizes the dsRNA. Other experiments suggested that the discrimination of self from nonself depended on the structure of dsRNA rather than its length.

So, there were three important questions that these findings raised. Where did the dsRNAs come from? Was it Z-RNA that allows MDA5 to discriminate between host and viral dsRNA? And what would cause Z-RNA formation? The first question had been answered by studies from three labs published in different journals in 2004 within four months of each other. Interestingly, the time spread would have less if one journal had not been two months slower than the other two in putting one of the articles to press [67–70]. That's the way science often works: nothing, then suddenly everyone makes the same finding. The simultaneous nature of these publications likely was the result of chatter at meetings rather than truly independent discoveries. Only the principals know who heard what where and when. Probably everyone has a different recollection of how the events unfolded.

The labs all demonstrated that repetitive elements in the genome are involved in forming dsRNA. Not just any repetitive element, but a class of sequences called "Alu". These elements were named because they contain a d(AGCT) sequence recognized by an enzyme that was named after the species in which the enzyme was discovered (_Arthrobacter luteus_). When genomic DNA was cut with this endonuclease, a prominent band of around 140 base pairs appeared on agarose gels – the "Alu" band. The length of the fragment represents a single Alu repeat, but most Alu repeats are present in the genome as a dimer, representing a fusion of two Alus long ago in the past. The monomers in the dimer have a dA-rich spacer in between. These dimer Alu element are about 300 bases long and represents about 11% of the human genome [71]. The Alus spread by coopting proteins encoded by other repeat elements that copy their RNA into DNA while pasting it somewhere else in the genome. The Alu themselves do not encode any protein. That is why some think of these repeats as "junk" DNA.

Sometimes, a new Alu dimer is inserted close to an existing Alu dimer, but in the reverse direction. The two copies are referred to as an inverted repeat (IR). There were many different times during human history when Alu elements managed to massively invade the human genome. The different waves of attack produced Alu families of different ages close enough to each other to produce IRs. These Alus pose an existential threat as they can insert into active genes and disrupt their function. The attack was akin to saturation bombing of cities during WW 2. From this perspective, Alus are not just junk. Instead, they are dumb and dangerous, with no idea of the harm they cause [71].

When IRs are read out from the genome into a single strand of RNA, they pose another danger. Since the repeat sequences are inverted, the bases are complementary. They can fold together to form dsRNA that become substrates for editing by ADAR1. In fact, the three labs found that the majority of ADAR1 edits in a cell are to Alu IRs. It is from these dsRNAs that MDA5 filaments can form (Figure 6.2). In the embryos where ADAR1 was absent, these IRs bound MDA5 and activated the deadly interferon response.

So, the question was, do the inverted Alu repeats form Z-RNA? When I examined these sequences, it became clear that they contained a motif that looked like it could form Z-RNA [72]. I confirmed that this was likely by assessing how much energy would be necessary to perform the flip under physiological conditions. Rather than form Z-RNA strongly, the best interpretation was that these sequences were poised to flip if pushed. I called the region I identified the Z-box, potentially explaining why ADAR p150 would bind to Alu elements (Figure 6.3). The Z-box was indeed quite highly conserved in the Alu families (Figure 6.4) [73]. It likely promoted their transcription by RNA Polymerase III (yes, there is likely a Zα-related domain in this complex). It was surprising to me that no one else had made this finding, but not really: no one in the field thought Z-DNA or Z-RNA had anything to do with RNA editing.

So why not write a paper and address the question "What is the Biological Role of the Z-DNA Binding Domains in ADAR1?"? So, I did. I hadn't composed a manuscript for a while. I wasn't sure how it would be received at a scientific journal. I did not have an academic affiliation. No one had ever heard of InsideOutBio. Not surprisingly, the paper was rejected by a number of journals. A typical communication with an editor would go like this: "While the piece is clearly written and we are generally interested in the topics explored, there were concerns about their immediate interest to a broad audience, which precludes us from considering it further". (Ines Chen, email to AH, August 21, 2018). To which I made the obvious reply, "That was fast! All I can say is that everything new starts small. Without an audience, that's the way it will stay".

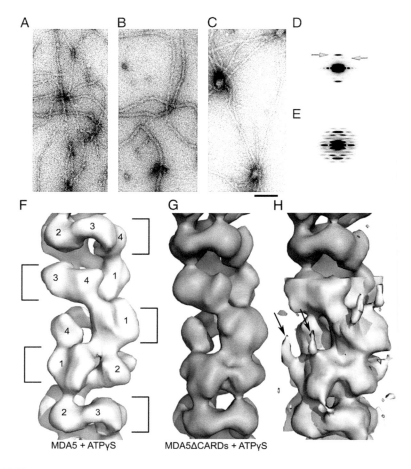

FIGURE 6.2   MDA5 filaments viewed by cryo-electron microscopy (A–C) and the structure calculated from the different projections of the fiber in the images (from *Proc Natl Acad Sci* 109, pp. 18437–41, 2012).

Ines Chang from Nature Structure Molecular Biology suggested I try their new journal *Communications Biology*. With the help of Dominique Morneau at the journal, the manuscript was published with the title "Z-DNA and Z-RNA in Human Disease". Dominique actually edited the paper – she spent many hours going through it (Microsoft Word time-stamped her edits) and made many helpful suggestions for improving the logical flow of the manuscript. The paper has now been accessed over 20,000 times and is cited over 130 times. I have since written similar articles that are either called reviews or perspectives. I think of them more as previews as they reflect an enormous amount from analyzing existing data to exclude sensible, but incorrect, explanations for the findings. In the process, I generate interesting questions worthy of further experimental study. It is always pleasing when the wet lab work does not disprove the conclusions drawn from the analysis. The challenge is to

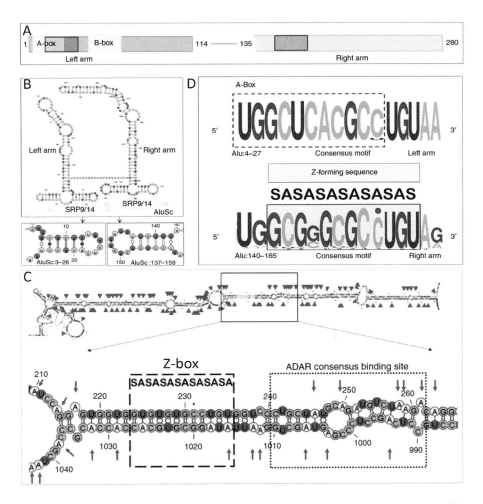

FIGURE 6.3 Alu inverted repeats form a fold-back structure that controls a Z-box. A. The linear representation of an Alu dimer that fuses two monomers. B. Each monomer of the Alu dimer can fold separately and they are joined by an adenosine-rich linker. C. A pair of Alu dimers that are on the same DNA strand but in reverse orientation can fold back on each other when transcribed into RNA to form a long double-stranded RNA that is an editing substrate for ADAR1. The A →I editing sites are indicated by arrows. The Z-box and the ADAR1 binding site are indicated within the dotted lines. D. The dinucleotide repeat of alternating *anti* and *syn* nucleotides are indicated (adapted from *Comm Biol*, 2019 Jan 7:2:7).

pose the questions in a way that the experimenter performs the necessary controls in an unbiased manner. Hopefully, other controls will be run as well to ensure that the result is robust.

The Alu Z-box prediction I made turned out well. Structural studies performed by Bert Vogeli and Quentin Vincens confirmed that the Zα domain binds to the Z-RNA conformation of the Alu IR sequence I illustrated in the *Communications*

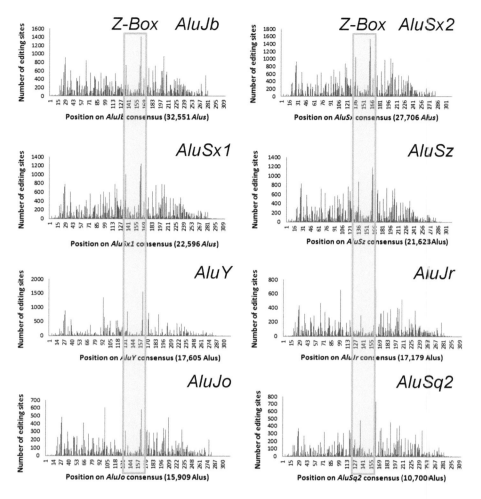

FIGURE 6.4 There are a number of Alu families that invaded the human genome at different times. The y-axis indicates the editing at different Alu dimer positions. The lines indicate the Z-box on the right monomer that is notable for the absence of edits. In contrast, the same region on the left monomer is edited (adapted from *Genome Res*, 24, pp. 365–76, 2014).

Biology paper. Their work also revealed more information concerning the role of non-Watson-Crick base pairs in the formation of Z-RNAs [74]. The presence of mismatches in dsRNA and loops and bulges allowed the flip from A-RNA to Z-RNA to occur at lower energies. There is no need to pry open the dsRNA to create a junction. There already exists a space in the base pairs which can invert. Their paper was a nice surprise!

Even with the experimental validation of Z-RNA formation by the Z-box, the ADAR1 editing community was not convinced. So, maybe they wanted to know what drove Z-RNA formation? The answer lay in the length difference between

A-RNA and Z-RNA helixes. The right-handed dsRNA is much shorter than the left-handed version – 24.6 Å versus 45.6 Å. Any stretching of A-RNA would favor the flip to Z-RNA as this would relieve the tension created (Figure 6.5) [3]. Stretching requires energy but from where does the juice come from? The safe answer in biology to this question is "The energy arises from hydrolysis of adenosine triphosphate (ATP)". This tiny molecule releases energy by breaking the bonds between two of the three phosphate atoms. The increase in entropy offsets the cost of doing business in the same way that the release of hostages is an exchange for one or more of the preferred outcomes. The question then becomes, "What is the enzyme that hydrolyzes ATP to cause the dsRNA to stretch?". Helicases are a good candidate. They are enzymes that unwind the dsRNA into single-stranded RNA (ssRNA), using ATP to energize the breaking of hydrogen bonds between the bases. There are many helicases in a cell, including MDA5, the enzyme Carl Walkley knocked out to save the mice who had an enzyme-dead form of ADAR1.

Why would a protein like MDA5 initiate an interferon response rather than unwind the dsRNA to produce single-stranded RNA? Normally, when MDA5 untwists the dsRNA, the enzyme changes its conformation, triggering the hydrolysis of bound ATP and the release of MDA5 from the RNA to begin another cycle. With longer dsRNAs, that cycle becomes more difficult to perform. In those cases, MDA5 binds to multiple sites on the dsRNA. There are regions in between where there are no free ends for MDA5 to untwist. As MDA5 clamps the strands together, it is not possible to prevent the dsRNA from reforming as soon as a single-stranded region is produced. The effort is futile. As a result, MDA5 ends up extending the filament, creating the scaffold for initiating the interferon response. However, the Z-box provides a safety switch. What happens is that MDA5 scrunches the dsRNA as it clamps on (Figure 6.6). In the region between the MDA5 patches, the dsRNA is stretched,

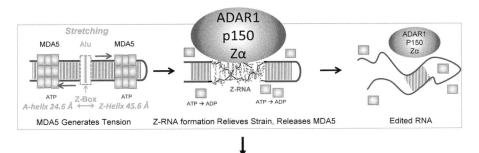

Suppression of Interferon Response Against Self RNAs
(but not against viruses)

FIGURE 6.5 Self-RNA have a repeat sequence that forms Z-RNA and allows ADAR1 p150 to turn off the interferon response that would otherwise be induced by MDA5. The filament MDA5 forms provide a scaffold for the proteins that induce the interferon-stimulated genes. Z-RNA formation by self-RNAs dissociates the filaments and prevents their reformation (adapted from *PLoS Genet*, 17:e1009513, 2021).

MDA5: ATP-bound   **MDA5:ADP-bound**

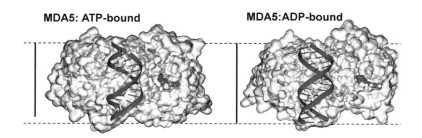

FIGURE 6.6 When dsRNA is bound by MDA5 loaded with ATP, the helix length is shortened (left panel). This creates tension in the dsRNA segments between the MDA5-bound patches. When the tension is released by Z-RNA formation, the ATP is hydrolyzed to form ADP (right panel). The black line to the left of each image shows that hydrolysis reflects the lengthening of the bound dsRNA (adapted from *PLoS Genet*, 17:e1009513, 2021).

creating tension. At some point, the Z-box will flip from A-RNA to Z-RNA. This will cause the dsRNA to lengthen: the Z-RNA helix is 45.6 Å long, while A-RNA is just 24.6 Å. The sudden relaxation of tension enables the change in MDA5 conformation necessary to trigger the hydrolysis of ATP and the release of MDA5. The Z-RNA provides a site for the p150 Zα domain to dock. The dsRNA domains of ADAR1 can then bind to prevent redocking of MDA5. The deaminase domain can subsequently edit the dsRNA and replace adenosine with inosine. The *coup de grâce* comes from enzymes that specifically degrade inosine-containing RNAs. They triage the edited self-dsRNA.

The interesting twist to this story is that Alu elements were once threats to the host genome. They now protect the host against attacking itself. ADAR1 p150 senses the formation of Z-RNAs by the Alu inverted repeats in host transcripts, terminating interferon responses that target only self-RNAs. The copy-and-paste sequence that once threatened the very existence of the host genome now has been tamed and repurposed to act as a guardian of the genome. The Alu repeats now mark host RNA and provide a landing place for the ADAR1 p150.

Of course, viral RNAs do not have Alu elements – they have no tolerance for such "junk". Viruses have found other ways to turn off the interferon responses that we will discuss in the next chapter. As of now, the mechanism of self-recognition based on Z-RNA formation by Alu inverted repeats is not well accepted by the RNA editing community. Yet, the mechanism is supported by multiple lines of experimental evidence: the formation of Z-RNA by the Z-box of IR is induced by Zα, cryo-electron microscopy shows the lengthening of dsRNA bound by the MDA5/ADP complex, the role of ATP hydrolysis in controlling the release of MDA5 is from a dsRNA filament (Figure 6.6); Alu IRs are pulled down with either anti-Z-RNA antibodies or by antibodies to the Zα domain; Alu inverted repeats are also present in interferon-stimulated genes, including those that code for MDA5 and another interferon response protein called PKR. There are other experimental proofs possible that eventually will be performed. Despite all the reluctance to embrace this model of self-/non-self-discrimination based on Z-RNA formation by Alu elements, the march of science is relentless. The best explanations for the data eventually are accepted, at least until a better one is found. So far, there is none.

During viral infection, Z-RNA can also form in tangles of cellular and viral RNAs. The dsRNAs arise when sequences that base pair with each other are in close proximity. Usually, the repeat sequences in RNA will nucleate the tangles by binding to any other matching repeat sequences. The tangles can also form when some step of viral replication is suboptimal, leading to the production of defective viral genomes that cannot undergo further processing by the viral replication and packaging machinery. As the RNAs twist, turn, and pair to form a double helix, there is sufficient force generated to flip segments of dsRNA to some other conformation. The tangles formed are no different from the bird's nest that arises during the casting of fishing tackle from a free-running reel. All is good until the line becomes entwined with itself. In the case of nucleic acids, the tangles stably trap many different alternative RNA folds, including Z-RNA.

The tangles can form from cellular RNAs that disengage from ribosomes when a cell is stressed. The RNAs, along with many proteins that bind them, form stress granules. Antibody studies show that the stress granules co-localize Z-RNA and the Zα protein domains that engage them. The role of stress granules is poorly understood. The old-school view is that these granules are anti-inflammatory and designed to protect cellular RNAs so that the cell can resume activities once the stress is removed. I suspect that, because stress granules can resemble viral replication factories, they are pro-inflammatory, even more so if the defective viral genomes stably fold into unusual RNA conformations. Binding of host sensors for these RNAs can then trigger an interferon response.

As I evaluated roles for Z-DNA in the biology of the cell, I searched out genetic studies to see whether amino acid variants of the Zα domain map to any particular phenotype. I had struck out on previous attempts while at the Framingham Heart Study and at Merck & Co. Not this time!

Genetic studies had previously linked ADAR1 to a Mendelian disease called Aicardi-Goutières syndrome (AGS) [75]. The system-wide inflammation is due to overproduction of interferon. One variant of the disease, known as bilateral striatal necrosis, causes calcification within the brain and early childhood death [76]. There were variants of ADAR1 that lacked the enzymatic activity necessary to edit dsRNA. Some cases required two different loss-of-function variants to produce disease, one inherited from each parent. Usually, the variant decreased ADAR1 enzymatic activity. Often, a variant, P193A, was involved. Here, the amino acid proline at position 193 in ADAR1 p150 was replaced by another amino acid called alanine (hence the designation P193A) (Figure 6.7). The proline at position 193 in the Zα domain was involved in Z-DNA binding and does not affect enzyme function. Markus Schade and I had previously shown that replacing the proline with alanine diminished the strength of interaction with Z-DNA. Structural studies performed using nuclear magnetic resonance by Markus and crystal structures by another student, Thomas Schwartz, confirmed that the P193 was essential for binding of Zα to Z-DNA. However, the mystery was that the P193A variant was also found in normal individuals without disease. In contrast, other ADAR1 variants causing AGS were

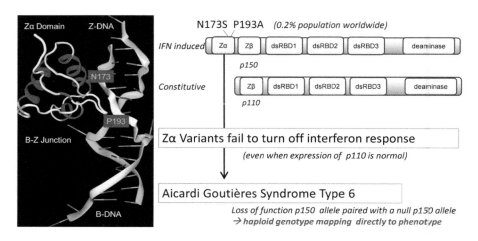

FIGURE 6.7 Zα variants cause the Mendelian disease Aicardi-Goutières syndrome type 6. Families with one null p150 allele and a loss-of-function p150 allele allowed the direct mapping of Zα variants to the Aicardi-Goutières phenotype (adapted from *Eur J Hum Genet*, 28, pp. 114-117, 2020).

only present in disease families. In fact, the P193A variant is quite frequent in the general population, at around 0.27% of ADAR1 alleles worldwide. That finding suggested that the P193A variant was not causal for disease. Maybe it wasn't?

Here is where the large amount of published data on AGS was helpful. I went over the information we had. We knew that p150 incorporates the Zα domain that is absent from p110 (Figure 6.7). We knew that expression of ADAR1 p150 was regulated differently from p110, requiring interferon for its expression. Due to this difference in the way these two protein products are regulated, expression of the P193A variant does not affect expression in normal cells of the p110 protein. Also, I found that some families had a variant chromosome that did not allow them to express p150 at all, i.e., they have a p150 null allele that made no p150 protein. However, they could still express p110 from the same chromosome. So, there was a wild-type allele that produces normal p150, a loss-of-function allele that produces P193A and also a p150 null allele that only made p110 protein. Additionally, I found that another Zα variant N173S (asparagine 173 to serine), which is also likely to diminish Z-DNA and Z-RNA binding by ADAR1 p150, was associated with disease. The three classes of Zα variants were what I needed to close the case [77].

I looked for families where both chromosomes express p110, but one of the p150 alleles was null (Figure 6.8). This situation meant that only a single chromosome expressed the p150 protein. What if the p150 expressed solely from this second chromosome was the loss-of-function P193A or N173S variant? Then we could find out whether P193A and N173S variants were directly causal for disease. In this situation, disease had to be due to the p150 variant as there was no normal copy of p150 to mask its effects. The disease could not be due to any problem with p110 or with the enzymatic domain as these were expressed normally in the affected individuals.

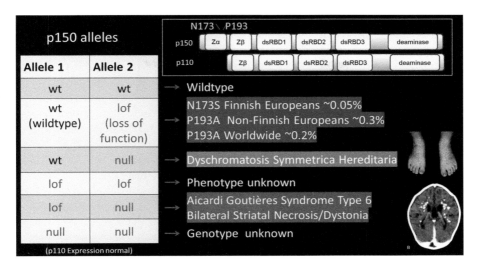

FIGURE 6.8 Phenotype of Zα variants showing the brownish macules on sun-exposed areas of skin in dyschromatosis symmetrica hereditaria (inset from Figure 1, *Brit J Dermatol*, 140, p. 492, 1999) and the brain calcification in severe forms of Aicardi-Goutières syndrome (inset from Figure 2, *Amer J Neurorad*, 30, p. 1973, 2009).

In fact, such families do exist with a P193A or N173S variant and a p150 null allele with normal p110 expression. The answer was clear. Either the loss-of-function P193A allele or the N173S allele is sufficient to cause disease when the only source of p150 in a cell. The Zα domain now had a phenotype. This finding was the first proof of a biological role for Z-DNA. I was really excited! I couldn't believe I finally had the result I had been looking for over the years.

The evidence just came together unexpectedly. I started writing the paper immediately. That did not take long as the words just flowed together. In fact, the figures took longer to prepare than the text. I sent the manuscript to *Nature Genetics*. Rejected without review. *BioRxiv* would not post the paper as it did not contain new research data. I pointed out that one of the papers from a former director of Cold Spring Harbor Laboratories, the home of *BioRxiv*, would not have qualified under such rules. I added to my response the link to the Watson and Crick *Nature* paper on the structure of DNA. My paper was reviewed and accepted by the *European Journal of Human Genetics*.

AGS caused by Zα variants occurs even though p110 levels are normal and the enzyme is still capable of editing double-stranded RNA. The genetics showed that the P193A variant no longer down-regulates interferon responses. Stated differently, this means that the Zα domain protects against interferonopathies induced by double-stranded RNAs. So why then is P193A so frequent in the world human population? The best guess is that the variant underwent selection during viral pandemics. The diminished p150 function allowed production of the higher interferon necessary to protect against viral spread. The variant is of highest frequency in non-Finnish

Europeans (0.3%). One idea is that the variant was selected during the rapid urbaniza-
tion that occurred during the Middle Ages where transmission of pathogens between
individuals was more likely. The measles virus is the most likely culprit. Normally,
victims die of measles because of secondary bacterial infection as the virus is so
efficient at suppressing immune responses. The P193A variant counters the virus by
permitting a more vigorous interferon response by the host. Interestingly, individuals
who have a P193A allele and a normal allele have increased risk of a skin ailment
called dyschromatosis symmetrica hereditaria 1 (DSH). DSH affects pigmentation
of the skin, but otherwise appears to produce no other serious outcomes [78]. The Zα
variants do not have a large impact on health when paired with a normal p150 allele.

My approach was based on a method called haplotype mapping. This technique
is commonly used in organisms like yeast that have only a single copy of each chromosome when in their vegetative growth phase. This analysis is usually not possible
in humans because we have two copies of every chromosome. I was lucky that the
way the ADAR1 gene was encoded generated a haploid state where the mapping of
Zα to phenotype was unambiguous. The result also highlighted the advances made
possible by sequencing the human genome and through the careful collection of
pedigrees by many highly skilled geneticists around the world. The findings have
been subsequently confirmed by introducing equivalent variants into the gene that
encodes ADAR1 in mice. Again, only when paired with a null allele did the Zα variant allele enhance the measures of interferon response.

It is interesting how the availability of vast amounts of data has changed the way
we do science, but not the way some scientists do science. After my paper describing
the genetics of Zα was published in the *European Journal of Human Genetics* [77],
I discovered, while doing a Google search, a meeting abstract describing early work
on the mouse version of P193A. I contacted Dan Stetson in whose lab the work was
happening and sent him my paper. I was curious to know whether he had looked at
the effects of P193A on editing. He responded that he would discuss further once the
full paper from his lab was in press. Eventually, a pre-print appeared on the *BioRxiv*
server with no mention of the relationship of the P193A variant to Z-DNA or Z-RNA
binding, and no reference to my paper in the *European Journal of Human Genetics*.
I was curious as to why. I emailed Dan to ask whether he had read the copy of my
manuscript that I sent him and he said he had. He did not cite it because it was a
"review". I guess he is correct in the sense that any new finding necessitates a review
of preexisting data to check for consistency. He seemed unable to accept that, in
this case, the human genetic analysis not only found the result faster than is possible
with mouse genetics, but was a method based on a synthesis of orthogonal data from
a wide variety of studies in different fields using unrelated approaches. With each
piece of evidence, the probability of a false discovery diminished. Old school versus
new school. As it transpired, Dan's paper relied on unreliable mouse mutant alleles
and his findings do not model the human disease [79, 80].

It was nice to have the genetic support to add to the structural and biochemical
elements and to provide an insight into the pathways involved. These results still
caught people by surprise. Especially coming from someone like me who seemed
to be out of touch with the real science everyone else was doing. I usually see

Phil Sharp, who co-discovered RNA splicing, once a year at the Koch Cancer Center Seminar at MIT. He knows so many people and he knows that he knows me so usually the conversation continues until he can place me. I did not waste any time telling him of the finding as it was still fresh and I was excited. "Hey Phil, we have a genetic phenotype for Z-DNA". (Phil) "What?" "Aicardi-Goutières syndrome". Pause. (Phil) "Send me the paper". Conversation over. Next year. "Hey Phil, we have the answer to what Z-DNA is doing in transcription". (Phil) "You don't have your name tag". I replied "They are only giving them this year to speakers". (Phil) "OK". Knowing the next line in the script, I said "I will send you the paper". (Phil) "OK". Conversation over. So it goes.

Nevertheless, the biology was unfolding fast. In a separate set of experiments by two different groups, cancer cells were screened to find genes critical for their growth of tumors in animals. The papers provided evidence that ADAR1 was important in cancer cell survival. This time, the focus was on how ADAR1 suppressed the interferon responses due to the dysregulated gene expression as malignancy progresses. ADAR1 allows tumors to silence the immune response (Figure 6.9A) in animals which are necessary to activate the body's defenses against these abnormal cells [81]. Just removing the p150 isoform was sufficient to produce tumor regression, again supporting a key role for the Zα domain in protecting the malignant cells [82]. Parallel work performed by a group of scientists in Cambridge, England on cultured cancer cells showed that up to 20–80% were dependent on ADAR1 (Figure 6.9 B) [83]. It's hard to know what is more surprising about this finding: the fact that ADAR1 was so necessary for tumors to survive or the failure by the authors of the paper to mention this role for ADAR1 as the data were buried in a supplementary spreadsheet. The reason for the lack of their comment was simple. The researchers were looking for DNA mutations that drove cell survival, not anything to do with RNA. You only see what you expect to see.

The reason for the dependence on ADAR1 is less surprising, given all the work on the genetics of interferon regulation by ADAR1. Although we are used to thinking in classical terms of the role for DNA mutation in causing cancer, there is also a need to consider other things that enable a tumor so obviously damaged and defective to survive the immune system. It is reasonable to believe that the abnormal proteins produced by this less-than-perfect cell should turn the body's defenses against it. But minds were not prepared for the misexpression of RNAs in tumor cells being able to also produce problems for the tumor. The abnormal amounts of dsRNAs produced in cancer cells would provide an inflammatory response that would drive the body's immune system to kill the tumor. By over-expressing ADAR1 , the tumors are able to silence the inflammatory response initiated by dsRNAs, especially those due to the repeat elements in the genome.

So, there we have it. A function for the Z-conformation based on Z-RNA! Yes. Unexpected? Yes. Exciting? Yes. New Biology? Yes. New Therapeutics? On the Way!

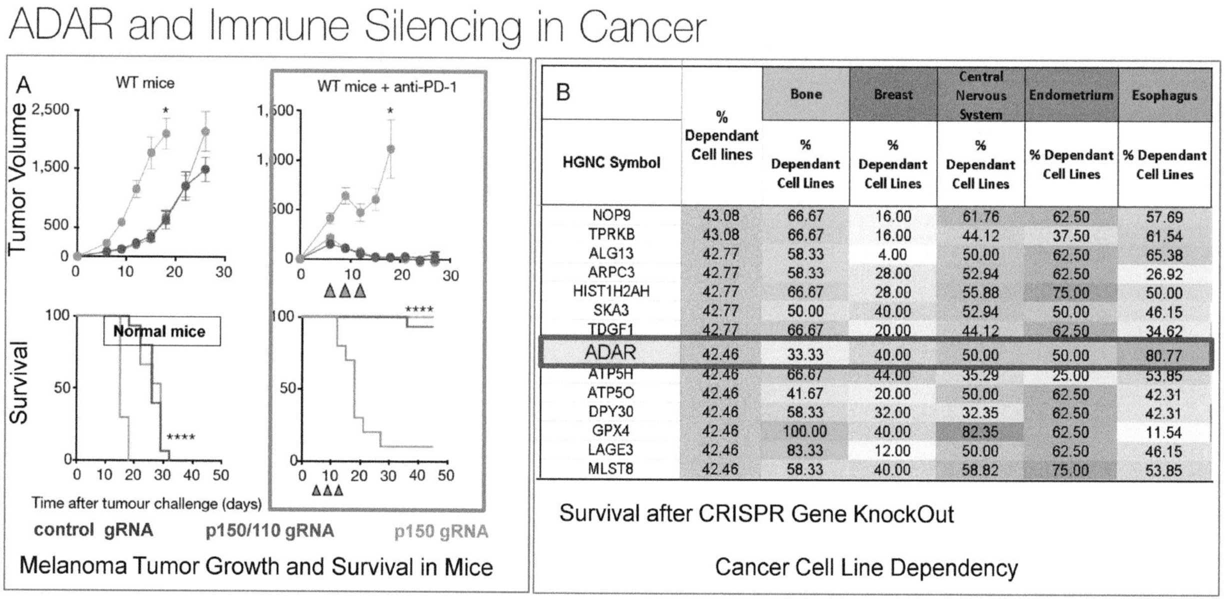

HGNC Symbol	% Dependant Cell lines	Bone % Dependant Cell Lines	Breast % Dependant Cell Lines	Central Nervous System % Dependant Cell Lines	Endometrium % Dependant Cell Lines	Esophagus % Dependant Cell Lines
NOP9	43.08	66.67	16.00	61.76	62.50	57.69
TPRKB	43.08	66.67	16.00	44.12	37.50	61.54
ALG13	42.77	58.33	4.00	50.00	62.50	65.38
ARPC3	42.77	58.33	28.00	52.94	62.50	26.92
HIST1H2AH	42.77	66.67	28.00	55.88	75.00	50.00
SKA3	42.77	50.00	40.00	52.94	50.00	46.15
TDGF1	42.77	66.67	20.00	44.12	62.50	34.62
ADAR	42.46	33.33	40.00	50.00	50.00	80.77
ATP5H	42.46	66.67	44.00	35.29	25.00	53.85
ATP5O	42.46	41.67	20.00	50.00	62.50	42.31
DPY30	42.46	58.33	32.00	32.35	62.50	42.31
GPX4	42.46	100.00	40.00	82.35	62.50	11.54
LAGE3	42.46	83.33	12.00	50.00	62.50	46.15
MLST8	42.46	58.33	40.00	58.82	75.00	53.85

FIGURE 6.9 Dependence of tumors on ADAR1 for survival in A. animal tumor models and B. cell cultures where CRISPR DNA editing using guide RNAs (gRNA) specific for p150 or both ADAR1 isoforms were used to knockout the ADAR gene, leading to the idea of immune silencing by the ADAR1 protein (adapted from *Nature*, 565, pp.43–8, 2019; *Nature* 568, pp. 511–16, 2019; *Trends Cancer*, 5, pp. 272–82, 2019).

7 Can Left-handed Z-DNA and Z-RNA Kill You?

When I identified Zα, we also identified a related protein in the database. At the time, there was not enough data to know that this was the second and only other Zα domain protein in the mammalian genome. I expressed the unknown protein as it had some interesting differences in sequence from the ADAR1 p150 and confirmed that this protein was Z-DNA binding. When the gene was cloned in 1999, the protein was named DLM-1 [84], then ZBP1 [85], then DAI [86] and then DAI/ZBP1/DLM-1 [87], then ZBP1/DAI [88], and finally ZBP1 again in 2020. The change in naming recapitulates the history of the field. But ZBP1 also binds Z-RNA, so it is misnamed (and why it is often called ZNA binding protein 1 where ZNA stands for both Z-RNA and Z-DNA). Another protein was also called ZBP1 (for zip-code binding protein 1) in 1994 by Rob Singer, so there was a period in the literature when the same name was used for two different proteins. There you have it: the non-science side of science where Z-DNA won out over zip-codes but ZNA would have been a better choice!

The complete protein sequence was pulled out of a screen in another laboratory in 1999 as an RNA message upregulated in macrophages, a cell type first named for its ability to eat things. The lab named the protein DLM-1 as it was isolated through the <u>D</u>ifferential expression of its RNA in the <u>M</u>esenteric stroma of animals that had ovarian tumors growing in their abdomen [84]. The RNA levels were higher than in the normal mesentery, a structure that attaches the gut to the rest of the body. The RNA was induced by interferon and <u>L</u>ipopolysaccharide but not by tumor necrosis factor (TNF). The Z-DNA-binding domain of DLM-1 was crystallized by the student Thomas Schwartz [85], who renamed the protein Z-DNA-binding protein 1 (ZBP1). Earlier, he had wanted to give Zα a new name but was not successful.

Not much more happened in the world of ZBP1 following its discovery in 1999 until 2007 when a paper in *Nature* from the Taniguchi laboratory proposed that ZBP1 was a cytoplasmic DNA sensor able to activate interferon-driven immune responses. They called the protein DAI (DNA-dependent activator of Interferon-regulatory factors) [86]. The next year, the laboratories of Ed Mocarski, working with Bill Kaiser, and Jürg Tschopp, working with Manuele Rebsamen, identified a domain in DAI that led to activation of a different inflammatory response pathway, one regulated by a protein called NF-kappa B [89, 90]. That response depended on a peptide motif that was called the RHIM (Receptor Homotypic Interacting Motif). This motif had been previously characterized in another protein that promoted death when a cell was exposed to TNF (Tumor Necrosis Factor) due to the activation of RIPK1 and RIPK3 (Receptor Interacting Protein Kinase). As the name implies, TNF was first discovered by its ability to kill cancer cells. The proteins released during

DOI: 10.1201/9781003463535-8

95

cell death then fire up the immune system to attack the malignant cells. This form of programmed cell death is different from other forms of cell death, such as apoptosis, that are designed to eliminate cells when they are no longer needed or after they have passed their "use by date". Then, in 2012, Ed, along with his first author Jason Upton, showed that interactions involving the DAI/ZBP1/DLM-1 RHIM were also capable of inducing the same form of necroptotic cell death as TNF. The difference was that DAI/ZBP1/DLM-1 acted inside the cell whereas TNF acted from the outside [87].

With these early reports, no one paid any attention to the two Zα domains in ZBP1 or to their functional significance! In 2016, a team led by Thirumala-Devi Kanneganti identified a key role for ZBP1/DAI in the cell necroptosis induced by the influenza virus. Their claim was that the Zα domains of ZBP1/DAI did not bind nucleic acids but responded to an influenza A virus (IAV) protein instead. They stated: "Our study demonstrates ZBP1 as an innate sensor of IAV proteins regulating antiviral innate immune responses" [88]. They had the story wrong. Two months later, Roshan Thapa and Sid Balachandran demonstrated that binding of DAI to RNA was sufficient to activate cell death during influenza viral infection [91]. The team introduced the equivalent Zα mutation into DAI that Markus Schade and I had shown was essential for Z-DNA binding by ADAR1 Zα [44]. The importance of the contact had been confirmed by subsequent crystal structures. Then, in 2020, Ting Zhang in the Balachandran group showed the RNAs bound by ZBP1 (no longer called DAI) were indeed Z-RNA [92]. The work established that sensing of Z-RNA initiated immune responses against viral infections. The earlier work on the function of RIP domains in Herpes virus infections could then be retrospectively interpreted as being due to the activation of ZBP1 by left-handed nucleic acids. Since Herpes viruses have a DNA genome that is very prone to form Z-DNA, it was uncertain whether Z-DNA, or Z-RNA, or both were being sensed. However, Jonathan Maelfait, working in Jan Rehwinkel's lab, and Haripriya Sridharan, in Jason Upton's lab, revealed that RNA transcription from the viral genome, but not DNA replication, was crucial for inducing ZBP1-dependent necroptosis [93, 94].

Unlike ADAR1, knockout of ZBP1 had no phenotypic effects by itself. There were a number of proteins that provided the checks and balances necessary to restrain this potentially dangerous protein that otherwise induces inflammatory cell death. The ZBP1-dependent pathways only became active in animal models when other regulators of necroptosis were genetically inactivated. Manolis Pasparakis and his laboratory showed that ZBP1 was capable of activating skin necrosis and inflammatory bowel disease in these genetically modified animals. The results indicated that endogenous cellular double-stranded RNAs were capable of forming Z-RNA to activate ZBP1. In situations when ADAR1 was unable to suppress Z-RNA levels in a cell, ZBP1 provided a fail-safe mechanism to eliminate cells that were no longer healthy. These findings caused a number of laboratories working on cell death and virus infection to enter the ZBP1 field (Figure 7.1). New roles for ZBP1 then emerged. ZBP1-dependent responses are triggered by damage to the telomeres present at the end of chromosomes and to the mitochondria that power cells [95, 96]. Altered telomere length and malfunctioning mitochondria are both hallmarks of aging.

As the work with viruses revealed, pathogens that induce Z-DNA or Z-RNA are rapidly destroyed through ZBP1-dependent pathways. Microbes and viruses protect

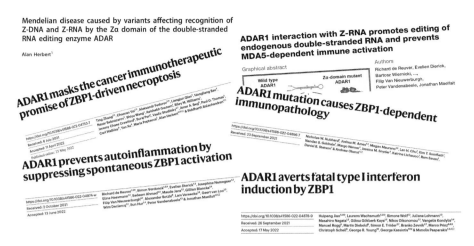

FIGURE 7.1   The field takes off with many high-profile publications.

themselves by inhibiting the ZBP1-initiated response. One example is the E3 protein from vaccinia virus, a relative of the variola virus that causes smallpox. I identified this protein as belonging to the Zα family in the 1997 *PNAS* paper. I visited Bert Jacobs in Arizona, who was working on the vaccinia virus, to initiate a collaboration with studies to be done in his lab and at MIT. E3 was not a strong Z binder and only flipped a B-DNA methylated d(GG)$_n$ polymer to the Z-DNA conformation, after a long incubation period (Figure 4.9). The delay enabled E3 to correctly align the key tyrosine involved in Z-DNA-specific recognition and then capture Z-DNA as it forms in the Z-prone methylated polymer. It was twenty-four years later, in 2021, that Heather Koehler and Ed Marcoski confirmed that vaccinia used E3 to suppress Z-RNA-dependent cell death responses [97].

Smallpox caused by the variola virus is a disease that has likely shaped the evolution of ZBP1. It is thought that up to 60 million people in Europe died of smallpox in the eighteenth century alone when a virulent strain first emerged. The pandemic caused up to 400,000 deaths per year (according to Wikipedia) and left the few survivors severely disfigured with scars on their faces and bodies. The battle between host and pathogen was fought with dueling Zα domains. The virus fought to prevent ZBP1 from killing infected cells. In response, ZBP1 unleashed the full power of the immune system, producing huge amounts of collateral damage that was patched up by whatever normal tissue survived the assault.

Many survivors were blinded by the smallpox virus. As Teresa Brandt and Bert Jacobs showed in animals infected with the vaccinia virus, the Zα domain in E3 enables infection of neural tissues. Another recent example where a Zα domain is essential for virulence is found in a different class of virus belonging to the Asfarviridae family [98]. That virus causes African Swine fever, producing encephalitis, ocular disease, pneumonia, and reproductive failure. The mortality approaches 100%. The virus is a commercially devastating disease for farmers as there is no treatment other than culling infected herds. Deletion of the Zα domain renders the virus harmless and is a strategy currently under evaluation for the production of an attenuated vaccine against the virus. Many years earlier, a similar strategy had been advocated for a smallpox virus vaccine but was never tested.

The studies provide evidence for a different role for ZBP1 in the immune response than the one played by ADAR1. ADAR1 turns off host responses by recognizing Alu sequences in host transcripts. In contrast, ZBP1 is activated by Z-DNA and Z-RNA to promote cell death. It is intriguing that the ADAR1 loss-of-function Zα variants in humans predispose individuals to the neurological disease that occurs in Aicardi-Goutières syndrome, type 6. Interestingly, work performed by Paul Marshall [99] in Australia revealed that knock out of either ADAR1 Z-DNA binding or editing diminishes new memory formation in mice. In particular, extinction of fear responses is greatly reduced as synapse formation is impaired. My best guess is that the underlying response initially arose to prevent viral spread across the synapse during infection. A protein remnant of an ancient retroviral gag protein called Arc is likely key to this outcome as Arc is still able to form capsids and ferry host RNAs across synapses. In its modern garb, Arc has evolved into a key regulator of neuro-plasticity though the RNAs it transports to the downstream neuron. During viral infection, Alu elements may be the major cargo transferred by Arc, especially when ADAR1 fails to suppress the expression of these retroelements. The Alu fragments then inhibit viral replication by repressing translation in the recipient neuron and impair synapse formation by shutting down new protein synthesis. The decreased memory consolidation when Alu levels are high likely contributes to the brain fog of long COVID infection and to the neurodegeneration in Alzheimer's disease.

The involvement of ZBP1 in these outcomes is currently unknown. One possibility is that ZBP1 shutdowns protein translation rather than kill neurons, mirroring the role of a Zα protein in gold fish called PKZ that limits viral infection by preventing their RNAs from engaging ribosomes and by promoting apoptosis, a non-inflammatory form of cell death. More generally, a stress response induced by the Alu elements may halt protein synthesis. Stress granules then form from all the discarded mRNA. The Z-DNA arising from the mRNAs tangles lead to ZBP1 activation, the enhancement of autophagy and the elimination of virus (see Figure 12.3).

Out of my conversations with Sid Balachandran about ZBP1 and Z-RNA, a collaboration grew that has become quite productive. We were both curious about how ADAR1 and ZBP1 interacted. The question was quite straightforward. How are the activities of these two proteins balanced? Although they have Zα domains in common, they are different in every other way. One protects self against self, whereas the other, if left unchecked, would kill everything. The two proteins are not twins, one good and the other bad. They each act when the other one fails, but both fail when one dominates. Maria Poptsova also joined our collaboration. Maria heads a very talented bioinformatics group in Moscow that has a strong interest in the role of alternative DNA conformations like Z-DNA in biology. Maria had contacted me after I reviewed her paper for an algorithm called DeepZ that was designed to find regions of Z-DNA in the genome. She felt the need to reach out to someone with experience in the field and was looking for an "international advisor" to help obtain funding for further work. I liked her approach and decided to explore the opportunity further. I also informed the editor of the journal about the contact as the manuscript had not been finally accepted. I wrote that "Dr. Poptsova approached me to help obtain a grant for her work on alternative DNA structures based on my publications in the area. I did identify that I had reviewed this paper to let her know of the potential conflict. I had not seen any revisions until now, nor discussed any specifics of the paper with her. The proposed work in Moscow relates to Z-DNA and RNA editing – a subject not covered in this manuscript – mainly because the Alu sequences of interest are routinely removed at the initial stages of RNA-Seq processing. I don't think this interaction has affected my views on the current manuscript – as you can tell from the review process, I raised issues that I believe have been properly addressed by Dr. Poptsova's team" (September 17, 2020).

We were beginning to build a team with a range of different skills and all interested in the same problem. Sid ran the wet lab and Maria the computational side of things. I was able to translate the findings from one realm to another and help frame the hypotheses we evaluated. Over the past few years, we have maintained our focus on the mysteries of Z-DNA and Z-RNA despite the COVID-19 pandemic and the dispersal of the group across the world in response to the crisis in Ukraine and the hopelessness of these events. One of our master's students Alex Fedorov has a co-first authorship on a *Nature* paper (his first manuscript) and is about to enter the PhD program at Oxford with Jan Rehwinkel as his mentor. Although we have been working together for quite a while, none of us has ever met personally. A new world order empowered by Zoom! Sid points out that 16 of the 19 authors on our Nature paper were born outside the US, illustrating again that good science knows no boundaries.

So, how do ADAR1 and ZBP1 interact? Sid had all the ZBP1 assays on tap. As with ADAR1, we could follow ZNA-dependent outcomes by the covalent modification of substrates in each pathway. Covalent in this sense means stable and long-lasting. In the case of RNA editing, it is the conversion of adenosine to inosine. The change persists in contrast to the transient nature of Z-RNA formation that triggers the modification. Whereas Z-RNA can be fleeting, there is the ADAR1 signature of "I was here" (I meaning the enzyme and also the edited base inosine). Of course, we proved that localization happened by mutating the residues in $Z\alpha$ that are key to Z-RNA recognition. The editing then no longer occurs, showing that the modification is dependent on engagement of the left-handed conformation. A similar approach can be used for ZBP1. We follow the phosphorylation events that depend on the interaction of ZBP1 with RIPK3 (Receptor Interacting Serine/Threonine Kinase 3), an enzyme that adds phosphate groups to its substrates. Again, we can mutate the $Z\alpha$ domains to show that the outcome is dependent on Z-RNA or Z-DNA. Ting Zhang in Sid's lab also developed a protocol to directly visualize Z-RNA and Z-DNA in cells. He uses the same monoclonal ZNA-specific antibody called Z22, developed by Eileen Lafer and David Stollar at Tufts, that I had used during the validation of my assay for the discovery of Z-DNA-binding proteins.

The problem Ting solved that limited the use of this antibody was a long-standing one. From the early days, it was possible to detect Z-DNA in cells with the Z-DNA-specific antibody [92]. Even better, you could light up bands in chromosomes that were sites of active transcription, showing that the energy for Z-DNA formation was likely generated as the RNA polymerases produced transcripts from the DNA template [100, 101]. All as expected. Then, Ron Hill, working with David Stollar, processed the chromosomes differently prior to staining, avoiding the acid fixation step used previously [102]. The bands detected were then the opposite of the previous finding. The regions stained with this different protocol were not transcriptionally active. The contradictory results depended on what the experimenter did and not on what was happening in the cell. There were also experiments published where Z-RNA was detected in small free-living creatures called protozoa, but the results of those studies cannot be validated as the reagents do not seem to exist anymore [103]. Fortunately, Z22 detects both Z-DNA and Z-RNA. Ting could distinguish the source of the staining by using nucleases that could remove either DNA, RNA, or a hybrid of DNA bound to RNA. If the staining was lost with one particular nuclease, then the source of the signal was established as either RNA, DNA, or both. We could validate the staining by showing that, under the conditions used, both ADAR1 and ZBP1 were activated in a Zα-dependent manner.

We began our investigations of how ADAR1 and ZBP1 interacted using the protocol Ting developed. One of our first findings was that Z-RNA was not detectable in cells with the normal ADAR1 p150 protein present. However, when the Zα domain in ADAR1 was mutated, Z-RNA appeared [104]. The Z-RNA accumulated over time and activated ZBP1 to kill the cell. Further, A-form dsRNA accumulated and set off the interferon response, driving the production of more ZBP1, MDA5, and PKR. The positive feedback loop finally broke the cell, causing rupture of cellular membranes. We showed that other RNAs also induced by interferon were pulled down by the Z22 antibody and by ZBP1, showing that they were capable of forming Z-RNA. They were not pulled down by a version of ZBP1 lacking the Zα domains. Editing of these RNAs depended on ADAR p150, as previously reported by Cyril George working with Chuck Samuels [105]. Curiously, the effects of Zα mutants in these earlier studies were not reported. Along with Yong Liu, who worked with Chuck, we had published a paper together in 1998 on the different domains of ADAR1 p150 to evaluate their functions, but the work stopped there [106].

Ting Zhang also performed a limited screen of molecules approved by the FDA for use in the clinic as cancer therapeutics (Figure 7.2). He wanted to see whether any of these drugs would induce Z-DNA or Z-RNA formation in cells. It was not obvious that any of the drugs would be useful to induce cancer cell death as many malignant cells have mutations that inactivate the ZBP1-dependent necroptosis pathway. The malignant cells avoid suicide by decommissioning the proteins that would trigger their demise. However, an analysis of tumors from mice revealed that the normal cells making up the stroma still had the ZBP1 cell-dependent pathway intact. Indeed, all three components of the pathway were expressed in cancer-associated fibroblasts (Figure 7.3).They were expressing the proteins because of the interferon induced by the presence of the cancer cells. But why was the pathway present in these cells not activated? The most probable answer was that the levels of Z-DNA or Z-RNA were too low to switch the response on. It was clear that ADAR1 expression was high enough to suppress the activation of ZBP1. That gave an explanation for why tumors express high levels of ADAR1. So, could we find a drug to increase the amount of ZNA in the tumor stroma to counter the actions of ADAR1? If so, the drug could kill off cells that were feeding tumor growth.

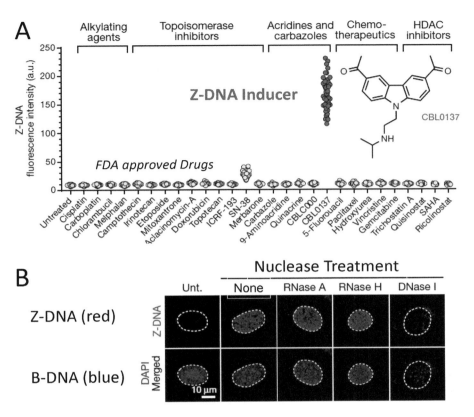

FIGURE 7.2 Discovery of a drug that induces Z-DNA in cells. A. After soaking clinically approved drugs into cells, Z-DNA formation was identified using the Z22 antibody developed by Eileen Lafer. B. The assay used with the antibody (adapted from *Nature*, 606, pp. 594–602, 2022).

Out of the screen came a drug, CBL0137, that induced ZNA. The compound had been tried in the clinic as a single agent with limited effectiveness. It came out of a screen for compounds that were targeting a completely different pathway. The drug had proven safe in a Phase I trial [107]. The low toxicity observed for the drug was important. Other approaches to activating tumor immunity based on B-DNA and A-RNA sensor pathways had failed because these sensors are always present in most normal cells. In contrast, the Z-RNA and Z-DNA sensors are not expressed under normal conditions. They must be induced by an interferon response.

So, if the pathway is active in the tumor stroma, and this drug induces Z-DNA, why has it not worked so far in the clinic? What were we missing? Previous work on the immunotherapy of cancer gave a hint. The problem is that tumors can block an immune response at many stages. Most importantly, the tumors express proteins that prevent amplification of an immune response. The anti-tumor immune cells are induced but do not proliferate. The introduction of the PD-1 antibody into the clinic overcame one of these immune checkpoints and was a tremendous triumph in the

Death Pathway is Expressed in Fibroblasts   Fibroblast Death is Activated by Z-DNA

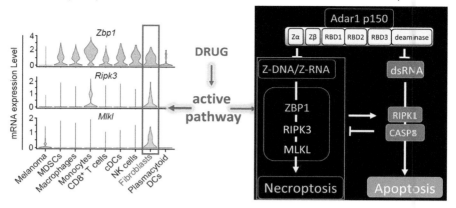

FIGURE 7.3   Killing tumors. A. The components of the pathway needed for Z-DNA to induce cell death are expressed in cancer-associated fibroblasts but not in tumor cells. B. In fibroblasts, the activation of ZBP1 by Z-DNA and Z-RNA can be suppressed by ADAR1 p150, preventing the use of the inflammatory cell death pathway. Editing of double-stranded RNA by ADAR1 can also suppress cell death through other pathways that dispose of cells more quietly, such as apoptosis. The drug overcomes ADAR1 inhibition by inducing sufficient Z-DNA to trigger the cell death of fibroblasts (adapted from *Nature*, 606, pp. 594–602. 2022).

treatment of cancer. Did we therefore need to combine CBL0137 with one of these immune checkpoint blockade (ICB) antibodies?

Our first step was to try a tumor that had not responded well to ICB in the currently employed mouse preclinical models (Figure 7.4). The combination worked and the tumor regressed. We could then use a mouse line in which the ZBP1 gene was deleted. In the absence of ZBP1, tumor growth was not affected by treatment with the drug and an ICB. We saw the same effect in a different tumor. A number of assays gave evidence that we were inducing a T-cell response against the tumor. The treatment caused regression of a second tumor in the animal that had not been injected with CBL0137, the so-called abscopal effect. Also, if we used a tumor expressing chicken ovalbumin, to which the mice had had no previous exposure, we could induce a T cell response against the ovalbumin peptides. These results gave proof that the treatment did induce a specific immune response and likely would do so against the abnormal proteins produced by a cancer cell.

It is exciting to take the basic science all the way to the clinic. We had potentially found a mechanism to bypass ADAR1 suppression of immune responses by tumors. By directly activating ZBP1 with a small molecule in the tumor stroma, we found a way to drive immune responses that kill tumor cells [108]. Most importantly, we can move from mouse studies directly to trials in humans as both CBL0137 and ICB are already in the clinic.

An open question now is how many other anti-cancer drugs work by disrupting the protection offered by ADAR1 to tumors. Many oncology drugs could act

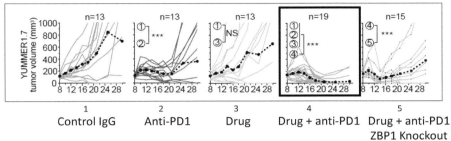

FIGURE 7.4 Treatment of a melanoma tumor with the drug CBL0137 and the immune checkpoint blocker anti-PD1 induces tumors to shrink in size, but not in animals where ZBP1 is absent (adapted from *Nature*, 606, pp. 594–602, 2022).

by overwhelming ADAR1 by further dysregulating RNA transcription in tumors. Others may change the localization of ADAR1 so that it is in the nucleus rather than in the cytoplasm where the double-stranded RNA sensors that activate immune responses are found. These possibilities are quite intriguing. They raise a number of questions not even imagined less than three years ago.

We are not done yet. We are still working with Sid to elucidate other ligands embedded within the different parts of the genome that protect against viruses and cancers. The work shows that the biology of Z-DNA and Z-RNA differs greatly from Z-RNA and B-DNA. The bottom line is that B-DNA and A-RNA sensors are present in every cell. The ZNA sensors ADAR p150 and ZBP1 are present only at times of inflammation or cellular stress. This differenceprovides many new therapeutic opportunities to target diseased cells while sparing normal tissues (Figure 7.5) Other groups have also published similar findings in *Nature* and are also contributing greatly to our knowledge about this novel Z-dependent biology and the the potential applications of flipon therapeutics [109–112] (Figure 7.1).

Figure 7.5 Right-handed A-RNA and B-DNA elicit different biologies than Z-RNA and Z-DNA. This outcome occurs because the expression of ADAR1 p150 and ZBP1 is interferon-dependent and highest during inflammatory responses. Z-DNAs and Z-RNAs that activate these sensors are highly transcribed in virally infected and stressed cells. They arise most often from repeat elements that can lie in introns, untranslated exons or within regions lacking any gene at all.

So, is Z-DNA bad for you? Z-DNA or Z-RNA is not there to kill you. These alternative nucleic acid structures protect you against pathogens and cells that are no longer functioning properly. Better a dead cell than one that is stressed beyond its limits. The ZNA-dependent responses exploit the suicide switches wired into cells. Cells constantly check how well they are doing. If things are going badly, the cell is programmed to take the next exit to nowhere. That act reflects a cell's focus on itself. In this sense, cells are quite introspective. They respond to their own responses. When they are not performing well, they react badly. This strategy allows them to sense threats by the levels of Z-RNA and Z-DNA present inside themselves and to

Immune Sensor	Normal Cell	Inflamed Cell
A-RNA/B-DNA	+	+
Z-RNA/Z-DNA	-	+

FIGURE 7.5 Right-handed A-RNA and B-DNA elicit different biologies than Z-RNA and Z-DNA. This outcome occurs because the expression of ADAR1 p150 and ZBP1 is interferon-dependent and highest during inflammatory responses. Z-DNAs and Z-RNAs that activate these sensors are highly transcribed in virally infected and stressed cells. They arise most often from repeat elements that lie outside genes.

react appropriately. It enables cells to detect threats that they have never previously encountered. If the cell ignores these troubling signs and decides not to hit the self-destruct switch , other cells will do what that cell failed to do by itself. They will kill the violator in order to protect the host.

The findings provide a reason for why Z-DNAs and Z-RNAs persist in the genome. They are there because of positive selection as they enable survival of the host. Retroelements only form Z-DNA when a cell fails to sequester them within the heterochromatin compartment. They only form Z-RNA when transcription becomes dysregulated. They lie in wait for viruses to grant RNA polymerases access to the regions beyond a gene's normal stop site. Both herpes simplex and influenza viral infections exemplify this outcome. The viruses disrupt the normal termination of trascription. Instead of stopping, the RNA polymerase continues making RNA, transcribing the Z-RNA-forming elements placed strategically to sense this type of unscheduled event. The Z-RNAs alert the cell that it is compromised and, when detected by ZBP1, activates cell death pathways, terminating the threat. The mechanism is simple but very general. The Z-RNA trap works against the current crop of viruses and against new ones that may emerge sometime in the future.

Surprisingly, Z-DNA plays a completely different role in protecting against pathogens. This time, ZNA formation is outside the cell where the threat is from the bacteria that live within us. Our intimate neighbors inhabit our skin, our bowels, our nostrils, and other places more private. Mostly, we coexist peacefully with our bacterial flora. It is only when the bacteria breach our barrier layers that we need to actively defend against them. Surprisingly, Z-DNA plays an important part in separating the host and bacteria from each other when this happens. The discovery by John Buzzo and Steve Goodman that Z-DNA was part of bacterial biofilms was unexpected [113]. By forming biofilms, bacteria are able to protect themselves against a host defense and also reduce their vulnerability to antibiotics, which kill them by weakening their cell walls They build an exoskeleton made of Z-DNA. Although cells make enzymes that cut up B-DNA with ease, Z-DNA is resistant to their action.

The bacteria build their Z-DNA exoskeleton with proteins on their cell surface that capture and bend right-handed DNA, torquing it sufficiently to flip it to the left-handed Z-DNA conformation. The binding of bacterial proteins at B–Z junctions decreases the overall energy cost of Z-DNA formation. Once tacked down, the Z-DNA is there to stay. From the host perspective, the Z-DNA exoskeleton encapsulates the bacteria, preventing their spread and allowing the immune system to contain the threat and, through abscess formation, eventually eliminate the invader.

Intriguingly, biofilms also undergo G4Q formation [114]. In the presence of hemin, a normal component of serum, the G4Q acts as an enzyme and can generate hydrogen peroxide in addition to that produced by the protein enzymes released from neutrophil granules [115]. Hydrogen peroxide is a highly bactericidal chemical species.

Of course, this strategy works well for the host in the short term but, in the longer term, there are risks. These include the development of anti-self-antibodies. Systemic lupus erythematosus is an example of the diseases that can result. While at medical school, it always fascinated me as to why there should be anti-nuclear antibodies in this disease. Indeed, the first Z-DNA-specific proteins discovered were the antibodies discovered by Eileen Lafer and Dave Stollar in the sera of patients with this disease. Their origin was a question that David Pisetsky at Duke was intrigued by, with the bacterial biofilm providing an answer to the source of the Z-DNA antigen. With biofilms, the long arrays of Z-DNA provide enough activation of B cells to stimulate an antibody response without the need for any help from T lymphocytes that normally drive an antibody response. Instead, the detection of bacterial products by myeloid cells stimulates sufficient cytokine production to drive the initial B-cell response. With time, the response is mediated by IgG_2 antibodies which are the predominant antibody class found in mice lacking mature T cells, due to a missing thymus. The response is also further amplified by neutrophils that also try and contain bacteria by enmeshing them in a DNA net. The net forms by expulsion by neutrophils of DNA into the extracellular space along with a protein called HMGB1 that can also bend DNA and promote Z-DNA formation. The inflammatory cycle can also break tolerance to other nuclear antigens, leading to the formation of immune complexes of antibody and antigen that can deposit in the capillaries of the skin and kidneys to produce inflammatory disease in these organs. The propensity to adverse outcomes is increased by a number of genetic variants that promote interferon production, B-cell proliferation and an overall failure to adequately clear immune complexes through opsonization by complement proteins. The disease is further exacerbated by viruses, such as the Epstein-Barr Virus, that promote the long-term survival of antigen-activated B cells [116, 117]. Therapies that diminish the pool of autoreactive B cells and that disrupt biofilm formation offer new approaches for the treatment of lupus disease flare-ups when coupled with appropriate antibiotics to contain bacterial infections. Intriguingly, there are bacterial enzymes like the staphylococcal S1 nuclease that will digest Z-DNA-containing biofilms, whereas human DNases will not [118].

This body of work and the many collaborations that made it possible established a biological role for left-handed DNAs and RNAs, explaining how Z-flipons are positively selected by viruses during evolution, linking the pathways involved to disease, and to new therapeutic strategies for their remediation. Not a bad comeback for a field once declared dead, a Z-phoenix arising from the ashes. Yet, this is just the first round of the Z-DNA comeback.

8 Does Z-DNA Regulate Transcription?

Of course, no one expected a role for Z-DNA in gene regulation, given the battle lines drawn up in the 1980s. Nevertheless, much progress has been made, even though the field was neglected for so long; during this period; everyone else was "eyes right", as they say in the military.

Initial investigations on the role of Z-DNA in transcription followed on from the work of Liu and Wang at Harvard [30]. In their "twin domain" model, a transcriptionally active RNA polymerase unwinds DNA in its wake, creating the conditions for a flip from B-DNA to Z-DNA in that negatively supercoiled domain. In the region ahead of the polymerase, the DNA becomes overwound to form a positively supercoiled domain. Their work was based on bacterial genomes. However, in the 1985 Jean L. Marx take-down of Z-DNA in *Science* (Figure 3.1), it was stated the "more recent results from Jim Wang's laboratory indicate that the plasmid inserts form Z-DNA inside the bacterial cells only under abnormal conditions".

The natural question to ask was, is the same result true of mouse and human genomes? Do they have sequences that flip conformation only under abnormal conditions? Or does the flip just happen routinely? If so, where is Z-DNA formed in the genome? Quite early on in the hunt for Z-flipons, algorithms were developed to search, chromosome by chromosome, for Z-DNA-forming sequences. The most basic approach was to look for sequences in regions where purines and pyrimidines alternated. That dinucleotide repeat pattern is expected from the zig-zag backbone first seen in the Z-DNA crystal structure. The sets of d(A-T) repeats were excluded from the analysis; even though this sequence is an alternating purine/pyrimidine repeat, it has a tendency to form other non-Z-DNA structures. The analytic approach was quite qualitative, just looking for yes/no pattern matches.

Later methods were quantitative and asked how much energy it would take to flip a given sequence into the Z-DNA conformation. The lower the energy, the better the sequence was at forming Z-DNA. In the Rich lab, these analyses were started by Mike Ellison and Shing Ho under the tutelage of Gary Quigley. Their aim was to locate the best Z-DNA-forming sequences in the genome [119]. Others also developed similar metrics, including work done earlier by Craig Benham [120]. These approaches suggested that the best Z-DNA-forming sequences were found in promoters and enhancers. Those gene elements assemble all the proteins necessary for an RNA polymerase to make a transcript from a gene. Promoters are normally very close to the site at which RNA synthesis starts (named, sensibly enough, as the "transcription start site" and abbreviated as TSS). Enhancers can be some distance away from the TSS. The proteins bound by an enhancer interact with those bound to a promoter to form a large assembly that is bridged by another set of proteins called

DOI: 10.1201/9781003463535-9

the mediator complex. The interactions may be quite extensive, with many different promoters bunching into the same region of the nucleus, forming a super enhancer to coordinate pathway gene expression and to specify cell identity. The mapping of potential Z-DNA-forming sequences supported a role for Z-DNA in gene regulation, but did not say why that would matter. Furthermore, there was no proof that the flip to Z-DNA ever occurred inside cells.

Studies in mouse and human cells, started in 1989 by Burghadt Wittig from the Freie Universität Berlin and later continued by Stefan Wolff, were designed to detect Z-DNA formation in genomic DNA inside cells [121, 122]. The cells were permeabilized with detergent to enable the diffusion into the nucleus of antibodies that recognized Z-DNA, where the genomic DNA was localized. Burghardt was able to show that Z-DNA was present in the nucleus under the conditions he used. The level of antibody binding could be increased with topoisomerase I inhibition, an enzyme that opposes Z-DNA formation by relaxing DNA regions that are underwound. That finding suggested that normally there was sufficient negative supercoiling in the nucleus to power the flip to Z-DNA. Burghardt observed that, with increasing amounts of antibody, there was a plateau region where only the same amount of antibody was bound despite adding increasing concentrations of antibody to the samples. This finding indicated that the Z-DNA was pre-existing and fixed in amount rather than induced by binding of the antibody. If Z-DNA was being induced by the antibody, the amount of DNA bound would increase proportionally to the amount of antibody being added.

In the initial system, it was difficult to determine how much the results were affected by the diffusion of proteins out of nuclei at the same time as the antibody was diffusing in. In such cases, loss of proteins that constrained negative supercoiling could promote Z-DNA formation. In later papers, the Wittig group demonstrated that the amount of Z-DNA antibody bound was determined by the number of actively transcribing RNA polymerases. In contrast, the DNA polymerases that replicated the DNA during cell division contributed little to the overall Z-DNA levels. The team then mapped Z-DNA-forming elements to the promoter of the MYC oncogene, variants of which commonly cause cancer. Interestingly, the degree of antibody binding diminished as cells were induced to develop into more mature cells. As differentiation occurred, the reduction in antibody binding correlated with the decrease in MYC gene expression. With this work completed, Burghardt then focused on more entrepreneurial pursuits. He remains enthusiastic about the role of dynamic DNA structures in biology.

Another approach for finding unusual flipon structures draws on earlier studies showing that the structure of non-B-DNA affects reactivity with base-specific chemicals. This field has a long history. The reactivity of some chemicals with DNA varies as the conformation of DNA changes whereas other compounds only modify DNA that is single-stranded. Brian Johnson, who I overlapped with in the Rich lab and who successfully managed a quick exit before the field imploded, extensively studied Z-DNA modifications by chemicals. Fedor Kouzine and David Leven took this approach and applied the methods to intact cells [123]. They used potassium permanganate ($KMnO_4$) to target thymines not hydrogen bonded with another base.

They could then identify the unpaired thymines associated with alternative DNA structures, particularly those present in B–Z junctions and others in the loops formed by G4-quadruplexes. They could also map single-stranded regions that arise in adenine- and thymine-rich duplex regions that have melted open under the stress of DNA unwinding produced by RNA polymerases. Fedor and David exposed cells to $KMnO_4$ for 60–90 seconds, providing a snapshot of the DNA conformations present at that moment in time (Figure 8.1). By matching the patterns observed with the

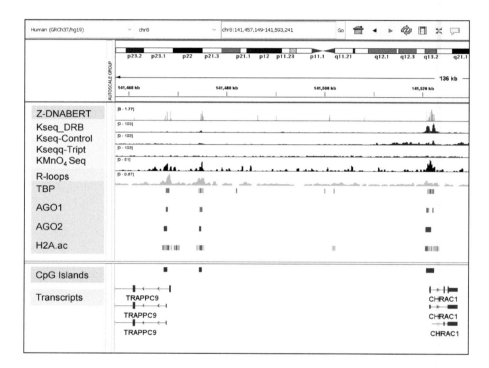

FIGURE 8.1 Mapping of Z-DNA to genomic locations. The Z-DNA-forming sequences are predicted based on chemical mapping of single-stranded regions containing guanines, using kethoxal (Kex), and thymines, using potassium permanganate ($KMnO_4$). Also, the locations of DNA-binding proteins, like TBP (TATA-binding protein), AGO (Argonaute) proteins, and histone proteins such as H2, that are activated by acetylation (H2A.ac), are determined by immunoprecipitation of the proteins cross-linked to DNA using specific antibodies, followed by sequencing of the bound DNA (chromatin immunoprecipitation-sequencing, ChIP-seq). Regions of RNA bound to DNA, that displace the other DNA strand (called R-loops) and gene transcripts, can also be determined by sequencing. The direction of gene transcription starts at the promoter region and proceeds in the direction shown by the arrows. CpG islands are sequences enriched for those dinucleotides. The mapping shows that Z-DNA is associated with promoters and that R-loops show enrichment there. DRB (5,6-dichloro-1-β-D-ribof uranosylbenzimidazole) is a drug that traps the RNA polymerase at a promoter, accentuating the Z-DNA signal, and Trip (Triptolide) prevents the polymerase from engaging the promoter to initiate Z-DNA formation

predictions of where flipons were located in the genome, they provided evidence that flipons do change their conformation under physiological conditions, confirming that flipons are active elements of the genome.

The experimentally determined flipons were a subset of all possible flipons, as the studies were performed on a very limited number of cell types under one or two conditions. With Dmitry Umerenkov and Maria Popstova, we were able to use deep learning based on the transformer algorithm to predict additional flipons genome-wide with an algorithm named Z-DNABERT (Figure 8.2) [124]. Again, we saw an enrichment in promoter sequences. Interestingly, we found a subset of variants that are causal for the Mendelian diseases that run in families which overlap with Z-flipons in around 3% of cases. These variants were often associated with short sequence insertions or deletions. The percentage increased to 9% if we looked for predicted loss-of-function variants that do not cause changes severe enough to be included in the mendelian disease database. Such variants are frequent enough to be found by sequencing DNA from a few thousand individuals.

We were also able to show an overlap with Z-DNA-forming sequences in the repeats within the genome associated with retroelements, especially long interspersed nuclear elements (LINEs) and endogenous retroviruses. These findings matched the enrichment of LINEs in the Z-DNA antibody pull-downs we saw with Sid and Ting in mouse cells treated with CBL0137 to induce Z-DNA formation. One possibility is that the cell expresses these elements as a sign of stress. For example, viruses often attack a cell by disrupting its production of mRNAs essential for normal function or injure the cell beyond repair. The cell may then respond to these disruptions by expressing a set of Z-RNA-forming sequences that are not transcribed in normal cells. These Z-RNAs then activate ZBP1 to induce cell death and eliminate the threat.

Interestingly, the Z-RNAs produced derive from retroelements that once invaded the human genome but now lie dormant in normal cells. These suppressed elements emerge from the shadows when the cell is losing its battle against a newer, more advanced interloper. Clearly, a game evolves in which the virus exploits a vulnerability to execute its host cell expeditiously and the host weaponizes that same vulnerability to eliminate an emergent threat. Attack and counter-attack. In some cases, it is better for the host to have a few cells die early rather than many die late.

But why have Z-DNA-forming sequences at promoters? The initial idea was that they bound sequence-specific transcription factors. However, there is currently no evidence for this model, though that does not mean that sequence-specific Z-DNA binding does not exist. The proposal seemed reasonable at the time as the information-rich, base-specific residues are exposed on the convex surface of Z-DNA; in B-DNA, they are buried in the larger of the two grooves that run around the helix. Currently, there are no hints as to the role that sequence-specific Z-DNA-binding proteins might play in the biology of a cell. That does not mean that recognition of Z-DNA by sequence-specific B-DNA-binding proteins does not occur. There may be a scanning of the exposed Z-DNA bases, followed by docking to the B-DNA conformers once the cognate sequence is found. No methods have been developed to look for this mechanism of "scan and secure".

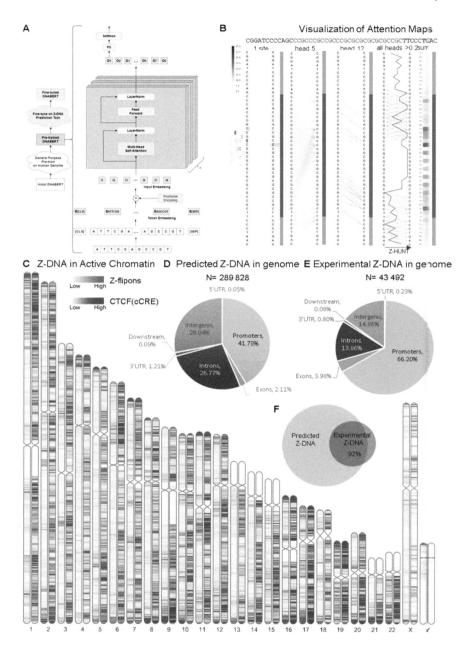

FIGURE 8.2 Genome-wide distribution of Z-flipons A. The different tokens and computational layers underlying the Z-DNABERT implementation. B. The transformer algorithm that processes experimental data through heads to find those features that best predict Z-DNA. C The chromosomal map of Z-DNA compared with that for an architectural protein called CTCF. D The mapping of Z-DNA to different genomic features and repeats (from *Life Sci Alliance*, 6, e202301962, 2023).

How else could flipons affect gene expression? The answer draws on the ability of flipons to store and release energy. The flip from B-DNA to Z-DNA enables the capture of chemical and mechanical energy generated as an RNA polymerase transcribes RNA. The chemical energy comes from the hydrolysis of the nucleotide building blocks that the enzyme couples together to form a transcript. The mechanical energy is from the stress arising when the polymerase unwinds the DNA to make RNA. As a Z-flipon changes conformation, it accumulates the energy released in the process of transcription.

The energy can then be used to reset the promoter (Figure 8.3). The reset requires removal of all the proteins necessary to load the polymerase onto the promoter. These proteins are normally tightly bound and do not otherwise come off the DNA easily. They must increase the DNA twist to open up the helix so that there is a bubble containing a region of single-stranded DNA for the polymerase to copy. The overwinding by the proteins generates positive supercoiling that also strengthens the interaction of protein with DNA. By reversing the positive supercoiling with the release of the negative supercoiling accumulated in Z-DNA, these proteins can be popped off [125]. The protein complexes then fall apart, allowing the cycle to start over again. The clearance takes a certain amount of energy to free the proteins. Z-DNA can act as an actuator. The ease with which the promoter resets depend on how many proteins are present in the promoter-binding complex and how much energy can accumulate in a particular Z-flipon. Altering either of these variables enables optimal tuning of the promoter reset. The process is analogous to the unraveling a fabric by pulling on a thread. The thread in this case is DNA and the weave is made with proteins. In contrast, suppression of Z-DNA formation can decrease gene expression. For example, Bimal and Alpana Ray found that, when the flip to Z-DNA in the ADAM-12 promoter was prevented, gene expression was lowered.

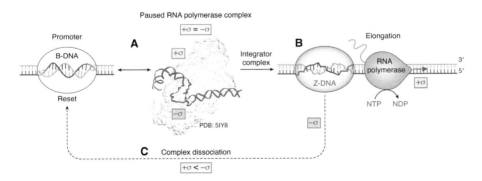

FIGURE 8.3 Binding of the RNA polymerase initiation complex (A) to the promoter generates the positive supercoiling $(+\sigma)$ that allows unwinding of DNA to open up the transcription bubble. The negative supercoiling $(-\sigma)$ resulting from RNA polymerase elongation is captured by Z-DNA (B). The release of the negative supercoiling offsets the positive supercoiling that stabilizes the complex, causing the complex to dissociate (C). Variations in the rate of dissociation and reassociation of the initiation complex allow regulation of gene transcription (from *J Biol Chem*, 299, pp. 105–140, 2023)

Z-formation may also be playing another role in maintaining high rates of gene transcription. The Z-DNA formed during the reset provides a mechanism to reinitiate binding of the transcriptional machinery. Indeed, the sequence and structural homology of factors like transcription factor E (TFE) in archaea with the Zα domain suggest that this protein may be Z-DNA binding, providing a direct link between Z-DNA and transcription. Interestingly, other components of the RNA polymerase machinery in humans show some evidence of relatedness. One of these is present in the RNA Polymerase III complex that can transcribe Alu repeat elements. The binding of Z-DNA by the POLR3C subunit would explain the persistence of the Z-Box in this class of retrotransposons.

Z-DNA formation and reset of the promoter can be modulated by methylation of cytosines. This modification lowers the energetic cost of flipping from B-DNA to Z-DNA. Modifications to proteins in the pre-initiation complex and to histones can further influence the strength of their interactions with DNA. The weaker interactions free DNA to flip conformation. Certain of these proteins may bend DNA to alter the ease with which B–Z junctions form. The flexibility at the junction contrasts with the overall rigidity of the DNA helix and relieves the tension if a bending force is applied to a DNA rod. Once the junction forms, adjacent Z-prone sequences can flip relatively easily as the energy cost will be quite low.

Negative supercoiling of DNA can also arise from the ejection of a nucleosome. These structures have a protein core made from two histone tetramers around which DNA is spooled (Figure 8.4). The wrapping allows storage of unused DNA in as small a volume as possible. At any particular time, s only a small fraction of the 2-meter-long length of DNA in each human cell is actively in use. Usually, unwrapping of the DNA occurs during the transcription of the RNA required for protein synthesis. At least, that is the traditional view of how nucleosomes function.

Nucleosomes, in fact, act like miniature batteries [126]. They store energy because the DNA wound around them is underwound. The nucleosome restrains the tension

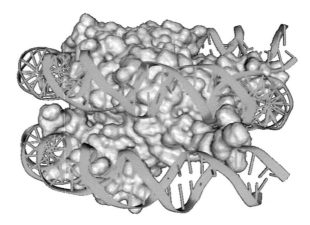

FIGURE 8.4 A DNA double helix wrapped around a nucleosome to form a compact structure that limits DNA transcription.

Part II

Here, we will discuss how soft-wired genomes assemble over time to make living things work. We will explore the alternative DNA structures, peptide patches, and the thermodynamic disequilibria that underlie their intransitive logic. We will find that no two cells are ever the same and that no two cells will ever respond identically. While no cell will respond perfectly in any and all situations, the response now, rather than later, decides a cell's fate.

10 What about Other Types of Flipons?

The name flipons is intended to capture intuitively the way a switch in DNA (or RNA) conformation can alter the programming of a cell. The name also follows the tradition where codons describe the triplet genetic code, exons refer to those expressed sequences found in proteins, and introns name the sequences within a gene that separates exons. The idea is that flipons are recognized by structure-specific proteins that assemble different cellular machines according to their conformation. They change the readout of information from the genome without a permanent change to the DNA sequence. They allow a cell to survive by exchanging energy for information.

One added benefit of the name flipons is that it is easy to explain the concept by handwaving. Of course, the handwaving does not imply flipons are a "maybe this or maybe that". Instead, the handwaving is quite informative. By inverting your palm, you can convey instantly that each side of your hand is shaped differently. One side can give a friendly handshake while the other side can make a threatening fist. Compare that intuitive explanation with the more technical alternative: "Well, DNA can be right-handed or left-handed … what that means is …". In Boston, the response to those few words is always rather rapid. It is usually goes something like, "How about those Red Sox?".

So far, we have focused on Z-DNA and Z-RNA-forming flipons. What about the other types of flipons? In science, one question leads to another. It is almost like you climb one peak only to see a higher one in the distance. It's like, "Damn, I thought we were done". On the other hand, the view from the next peak may be even better. You already have more experience and better climbing gear than you started with, so why not march onward?

Here is the story so far. The results that we have support the notion that Z-DNA and Z-RNA act as conformational switches to change cellular responses according to context. When in the right-handed conformation, the Z-flipon is bound by one set of factors, whereas, in the left-handed state, it is bound by different proteins. The interaction of ADAR1 p150 and ZBP1 with Z-RNA provide an example. The flipons involved turn immune responses on or off. They convey information by their structure and not by their specific sequence: just like the call of heads or tails does not depend on how a coin is etched.

There are plenty of binary switches in biology, so what is so special about flipons? For starters, they are encoded in DNA and are transmitted to offspring. As we will see, many binary switches, like the modifications made to proteins, are not templated. Next, Z-forming elements are quite frequent in the genome. They are well dispersed throughout active genes thanks to the waves of Alu invasion and the spread of repetitive sequences. Through variations in their conformation, flipons

DOI: 10.1201/9781003463535-12

allow different sets of genetic information to be read from the genome (Figure 1.10). Further, it is unlikely that all of the flipons in all of the genes are set in exactly the same way in all of the cells in your entire body. Stated more dramatically, flipons ensure that no two cells in an individual are ever the same [71]. Consequently, a cell will never read out the same information from the genome as any other cell and its phenotype will be unique.

In this sense, the genome offers an almost unlimited array of different outcomes. Selection both at the level of the cell and of the individual will determine which flipons are the most adaptive. The number of possible differences between siblings then is far greater than suggested by the random allocation of parental chromosomes each receives. On average, siblings will share a quarter of their genes ($\frac{1}{2}$ from each parent $=\frac{1}{2} \times \frac{1}{2} = \frac{1}{4}$). However, when viewed from the perspective of an individual chromosome, siblings have a $\frac{1}{2}$ chance of inheriting the same copy of chromosome 1 from their father, $\frac{1}{2}$ chance of inheriting the same chromosome 2 from their father, and so on. The same is true for the chromosomal copies they receive from their mother. Assuming random assortment, that comes out to be $(\frac{1}{2})^{23} \times (\frac{1}{2})^{23}$, a large number! The fact that we are basically quite similar is due to our common descent from a small group of ancestors who were prolific breeders. Much of our differences lie not in the genes we share, but in the repeat regions of our genome where flipons are most often found. The repeats are quite variable and are the basis of DNA tests that can confirm for you beyond a reasonable doubt that you are a 1 in a 9 billion type of person. The flipons also vary greatly in the way their conformation is set in various cells, meaning that there are many different versions of you that are possible (Figure 1.10). The particular "you" reading these words reflects the random events and adaptations that have set the current conformation of your genome to determine what information is read out in each of your cells.

Over the eons, the distribution of flipons in the genome varies as retroelements and sequence repeats spread through the genome through retrotransposition, recombination, and repair pathways. The outcomes will be subject to selection. Such processes provided an easy explanation for the nonrandom distribution of Z-DNA-forming sequences found throughout human chromosomes (Figure 8.2). However, the non-random distribution could instead reflect a particular set of flipons that our ancestors just happened to have for no particular reason at all. A different flipon distribution may have been found if a different set of progenitors managed to survive all the population bottlenecks caused by pandemics and other forms of adversity. We just don't know what their vanquished contemporaries of our ancestors had to offer in the way of flipons. However, since the location of flipons in genomes is similar in different populations across the world, it is likely that selection, rather than founder effects, accounts for the current flipon distribution we find in genomes. The founder effects mostly impact flipon length and sequence composition rather than location. Z-flipons do not appear to be just the fluff of junk-riddled genomes but rather a selected set.

Flipons are the makings of a digital genome. The programming can switch rapidly. Cells can reset flipons to optimize outcomes. No cell is likely to give the perfect response in any and all situations, but they need to respond now, not later. Cells then undergo selection so that tissues are populated by those that are the best adapted to the current contingencies. Depending on history, a subset of stem cells will populate the body organs while others will not contribute much. That is, unless circumstances change. The stem cell population that works best for a particular situation is chosen. Of course, you don't really need to get rid of any stem cell unless it is stressed or exhausted. You can always hold stem cells in reserve.

The principle of ongoing selection of somatic cells is illustrated by studies from Bevin Engelward's lab at MIT on how tissues evolve with age (Figure 10.1) [145]. Her team has shown that different cell clones emerge in the pancreas as animals grow older. Her group tracks DNA in the progeny of a particular stem cell by looking at DNA recombination events that lead to the expression of a fluorescent protein. Often, repeat elements are involved in these recombination events. Due to their high frequency in the genome, repeats at either end of the broken DNA strands can also help stick the pieces together, due to their sequence homology, and guide the recombination and repair process. As the lesions are resolved, the repeats themselves can grow or shrink and change the expression of nearby genes. Although often without consequence, adverse effects can arise when the repair triggers large-scale DNA rearrangements and inappropriately induces the expression of cancer-causing oncogenes or deletes tumor suppressor genes. Indeed, Bevin Engelward's technique requires the in-frame fusion of two gene segments to drive the production of the fluorescent protein she scores in her assays. What the results show is that there is somatic selection of cells and that different clones over time can emerge to occupy a significant part of a tissue. In principle, variations in gene expression associated with differences in flipon conformation can also provide a selective advantage for a subset of stem cells, the descendants of which can adapt the function of a tissue to the environmental exposure an animal experiences. These processes impact phenotype.

The changes due to repeat-associated DNA damage in somatic cells, which occur and are captured by Bevin Engelward's assay, arise at a one to two orders of magnitude higher frequency than do the transmissible genetic variations arising in germ

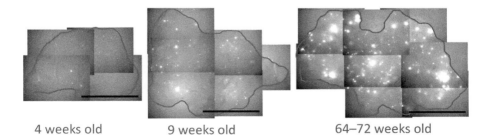

4 weeks old   9 weeks old   64–72 weeks old

FIGURE 10.1 Effect of aging on the number of DNA recombination events in a mouse pancreas. The white stars represent clones where DNA rearrangements have occurred. Laboratory mice live 112–130 weeks, while mice in the wild live about 17 weeks (adapted from *Proc Natl Acad Sci US*, 103, pp. 11862–7, 2006 (copyright (2006) National Academy of Sciences, USA).

cells. The somatic variants potentially contribute much of the trait value variation and common disease risk we see as people age. They are not easily mapped by the genome-wide association studies we discussed earlier where the genotyped DNA is mostly derived from blood. As shown by the GIANT consortium, just mapping single nucleotide variations in five million people, mostly Europeans, can at best account for half the measured heritable variation in height [146].

Flipons also likely contribute to the heritability which is not currently detected in genome-wide association studies. Their conformation varies with context. They affect trait values by changing the readout of genes without altering their DNA sequence or causing DNA damage.

There are many flipon types other those that form Z-DNA. They impact cell biology in different ways. In the next section, I will describe the flipon folds and then go on to explore the manner in which the different flipon classes shape heritable phenotypes.

OTHER TYPES OF FLIPONS

What other alternative DNA structures exist? There are many (Figure 10.2)? Should we call them flipons? Yes! They adopt different shapes with different properties and different effects. What is interesting is that these alternative nucleic acid conformations are also formed from simple sequence motifs. Most were discovered soon after Watson and Crick proposed their B-DNA model. These alternative folds were curiosities that were found once the first enzymes (generally called polymerases) were purified that were capable of catalyzing the synthesis of nucleic acid polymers. A surprise finding by the Ochoa group at the NIH was that not all enzymes capable of forming RNA polymers (polymerases) required a template. This meant that you could make long RNA strands of poly-adenosine (poly-A), poly-guanosine (poly-G), poly-uridine (poly-U), or poly-cytosine (poly-C). You could also make copolymers just by varying the mix of nucleotides given to these enzymes. Those insights were valuable in cracking the genetic code as the polymer sequences directed the incorporation by the ribosome of specific amino acids in a specific order into a peptide

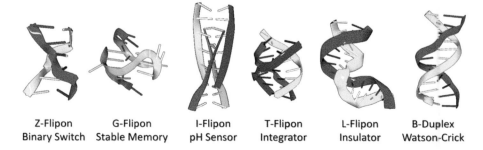

| Z-Flipon | G-Flipon | I-Flipon | T-Flipon | L-Flipon | B-Duplex |
| Binary Switch | Stable Memory | pH Sensor | Integrator | Insulator | Watson-Crick |

FIGURE 10.2 Examples of flipons that form different structures from two, three or four strands of a nucleic acid polymer. Possible functions are indicated below the name. The Watson-Crick B-DNA helix is on the right.

chain, revealing the underlying cipher. The utility of these template-independent polymerases offset the fact that they were different from the postulated Watson and Crick DNA replication polymerase that required a DNA strand to copy.

But then most of these first polymers were made of RNA. It soon became apparent that their repeat sequences could form three-stranded and four-stranded structures, in addition to the double helix. In 1957, Gary Felsenfeld, working with Alex Rich, performed careful mixing of poly-A and poly-U, and examined how the absorption of UV light depends upon whether the RNA is double-stranded or single-stranded. Their intention was to find a structure for RNA equivalent to that of the Watson-Crick DNA model. To their surprise, they found that poly-A mixed with twice the number of poly-U strands formed something different, with a unique X-ray diffraction pattern [147]. To Ochoa's surprise, according to Alex, the hybridization, as it is now known, did not require an enzyme. Discovery of the four-stranded structure was a little trickier. In 1958, Alex flying solo did produce an X-ray diffraction pattern of the poly-inosine quadruplex, but erroneously fitted a triplex model [148]. It was not until 1972 that Struther Arnott obtained, in his words, "somewhat better-quality diffraction patterns" (his circumlocution, [149]) and revealed that a quadruplex was the correct solution. Such is science – you give it your best shot with the cleanest data you can produce. You hope others forgive you your errors when the signal is subsumed by stochasticity and wish that they don't attribute it to your sloppiness as an experimentalist.

Intrinsically bent DNA was discovered in 1982 (called L-flipons in Figure 10.2 to reflect their shape). The structure was formed by homopolymeric dA-dT base pairs ("A tracts") and was found by the laboratories of Crothers and Englund [150]. Both groups were studying the kinetoplast body of the eukaryotic parasitic protozoan *Leishmania tarentolae*. This structure is bound by a class of high-mobility proteins, as detailed by a series of high-resolution crystal structures. An example is the center panel of Figure 10.2. Yet another structure based on poly-cytosine was discovered by Maurice Guéron in 1993 [151]. The structure is four-stranded, with one pair of cytosines stacked between the pair of cytosines above and below. The intercalation (one pair between another pair) of the cytosine base pairs gives rise to the I-motif name. Tom Jovin also provided evidence for the formation of duplexes with strands that are parallel in contrast to the anti-parallel strands of the Watson-Crick DNA model. Quite an exotic menagerie, indeed.

The shapes that DNA form are quite unique and differ greatly from the Watson-Crick B-DNA. In one way or another, they can form by flipping, bending, folding, or annealing the right-handed B-DNA helix without any strand breakage. Whether they all exist in cells is of great interest. It is likely that they all flip on one occasion or another. Their formation certainly can be powered by the energy released as a nucleosome is evicted from DNA. As the DNA uncoils, the strands can separate to allow the alternative conformations to fold (Figure 10.3).

The ease of with which each flipon class forms an alternative structure varies, as does the stability of the non-B structure. The flip to Z-DNA is quite dynamic. The ease with which a Z-flipon changes state depends on the repeat sequence involved and how readily the base pairs can invert to generate the alternating *syn/anti* zig-zag backbone conformation (Figure 1.1). The flip back to A-RNA or B-DNA is also quite

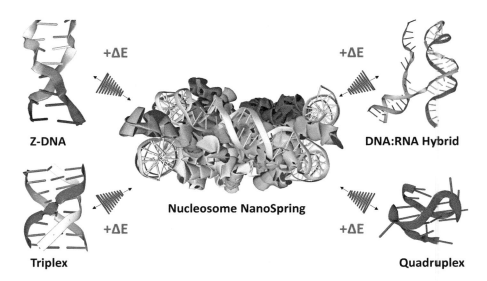

FIGURE 10.3 Energy is stored by wrapping DNA around a nucleosome, coiling the DNA like a spring. Ejection of the nucleosome releases energy to power the formation of alternative flipon conformations (adapted from *Bioessays*, 44, e2200166, 2022).

rapid as these are the lowest-energy configurations under physiological conditions. In contrast, other flipon classes require formation of a sufficiently long, single-stranded region to initiate the fold and can require more energy. The alternative structures formed can be quite low energy and persist for a much longer time than Z-DNA.

G-FLIPONS

G4-quadruplexes (G4Q) can be formed by single-stranded DNA and RNA in multiple ways. Depending on the loop sizes, the nucleic acid strands can be aligned into parallel, anti-parallel, or both parallel and anti-parallel arrangements. In each case, the structure is more compact than the unfolded strand and so they shorten the DNA (or RNA) as they form a knob-like protrusion. They cause the helix to kink or bend. There is an energy cost to forming DNA quadruplexes as it is necessary to break the hydrogen bonds that stabilize the B-DNA duplex. Once formed, the quadruplex itself is more stable than Watson-Crick DNA due to the way the bases stack one on top of another and the extra hydrogen bonds formed between the strands. Quadruplexes can also be formed in other ways from single stranded DNA (or RNA). Four single strands that are not otherwise connected to each other can align to form G4-wires. Two independent duplexes can also engage to form a G4Q. The duplexes then could zip homologous chromosomes together as the copy received from each parent will have the same spacing of quadruplex-forming sequences.

The distribution of G-flipons in the genome is also non-random. Much interest in G-flipons has arisen because G4Q-forming sequences are present at chromosomal ends and prevent chromosomes from fusing with one another. However, an alternative conformation, first proposed by Jack Griffith and Titia de Lange, is currently the favorite model in humans for how telomeres are protected [152]. In this scheme. the telomere ends fold back onto a duplex region and form a three-stranded structure with the single-stranded chromosomal tail.

Overall, the experimental studies are more complex than with Z-DNA as there is no binding domain equivalent to Zα that is specific for the G4Q structure. Instead, the recognition of G4Q may depend on the single-stranded loop regions created by the four-stranded fold or on the landing pad formed by the ends of a G4 stack. The crystal structures of a yeast protein, RAP1, shows the engagement of the G4 quadruplex through an α-helix that sits on top of the G4 stack. Intriguingly, the same α-helix is also used to bind B-DNA, with a different face of the helix engaging each structure [153]. G4Q is bound by the hydrophobic face that is usually buried in order to hide the residues from water, whereas B-DNA is bound by the positively charged face of the α-helix. This outcome means that it may be difficult to separate B-DNA-binding proteins and G4-binding proteins if they all use the same structural component for binding both conformations. Other peptides with positively charged patches of amino acids can also show preferential binding to G4, providing a way to bridge these structures with proteins that perform specific functions.

In many organisms, G4 formation is also regulated by a wide array of helicases that can unfold these shapes and revert them back to a single-stranded regions. The helicases first grab onto the single-stranded loops at the ends of the G4 stacks, then tease the structures apart [154]. Genetic mutations in helicases are associated with disease, showing that dynamic control of G4 dissolution is essential to avoid deleterious outcomes from arising. Some of the G4Q-resolving helicases are also involved in RNA splicing, although how G4Q is involved in such events is not fully established.

Another possible role of G4Qs is suggested by their enrichment in promoter regions, where they may regulate gene transcription by preventing an RNA polymerase from elongating a transcript until an appropriate complex is assembled at the site of G4Q formation. Such outcomes can also be regulated by RNA that transiently forms G4Qs to strip off repressive proteins from DNA, at least *in vitro*. We will discuss these mechanisms in more detail in the following chapters.

G-flipons can play other roles as they are very stable, compared with Z-flipons. They do not spontaneously revert to A-RNA or B-DNA as there is a large number of hydrogen bonds to break for the reset to occur. The detection of G4Q formation by cellular proteins can potentially trigger a number of pathways. During replication, G4Q can arise as the DNA strands are prized apart to copy. G4Q formation signals the need for helicases to dock and unfold the structure. In another situation, G4Q can flag DNA damage. The G4Q then localize repair complexes. G4Q can also form at promoters during transcription. These on the coding strand can then trigger the assembly of complexes that maintain the DNA in an open state to facilitate later rounds of RNA synthesis. It is even possible that G4Q can act as memory elements that record the transcriptional status of a particular cell and provide a way to transmit this information to its descendants.

G4Q may have some highly specialized functions. For example, in antibody-producing B cells, G4Q promote the switch of antibody class. The process involves changing the segment of DNA that is read out into RNA for form the constant region of an antibody. The switch occurs by replacing the old DNA segment that is adjacent to the antibody variable region with a new one. The order of change follows the order of the chromosomal DNA encoding the IgM > IgG3 > IgG1 > IgG2b > IgG2a antibody classes [155]. The class switch is directed by T cells and allows the targeting of antibodies to different surfaces or different cells, increasing the specificity of response to the ongoing threat.

Some parasites, such as the malarial *Plasmodium falciparum*, use a similar rearrangement mechanism to evade the host immune system. Here, the DNA flip switches the segment that is read out into RNA, evading the host antibody response by changing the surface proteins expressed by the parasite. While G4Q involvement in switching is uncertain, the structure plays an important role in generating novel surface antigens for the parasite to express. It is likely that the helicases tasked with resolving G4Qs structures promote DNA recombination between those segments that encode the surface protein to generate new variants that can then be switched in and out as needed [156].

T-FLIPONS

Three-stranded structures, called triplexes (Figure 10.4), are favored by a different type of flipon repeat, one in which purines are repeated many times over. The T-flipon sequences may be all "Gs" or all "As", but often they are a mix of these purine bases. In a few cases, alternative hydrogen-bonding schemes can allow inclusion of a modified pyrimidine base. The triplexes can form in a mirror repeat of DNA. Here, one DNA strand separates from the other and folds back into the major or minor grove of an adjacent B-DNA duplex with a sequence that matches when read in reverse The third DNA strand hydrogen bonds to the duplex DNA in a base-specific manner without disrupting the Watson-Crick base pairing. The interactions involve either a Hoogsteen or reverse-Hoogsteen hydrogen-bonding scheme that is distinct from the Watson-Crick base pairing arrangement. These structures are called H-DNA as they were initially found under acidic conditions (marked by an excess of positively charged hydrogen ions or H+). The fourth DNA strand remains unpaired in these situations.

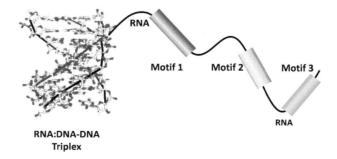

RNA:DNA-DNA
Triplex

FIGURE 10.4 A T-flipon with the third strand composed of an RNA. Appended to the RNA are additional motifs that dock proteins to help build a cellular machine with specific functions.

Triplexes can also form from a duplex and a single-stranded RNA produced at a different location in the genome, or even from a short piece of synthetic DNA or RNA that enters the cell from the outside. It is proposed that some of the RNAs that form triplexes actually do not code for anything. They belong to a class of RNAs called long noncoding RNAs (lncRNAs). Rather, the lncRNAs act as scaffolds to bind specific proteins and thereby carry them to a particular genomic location. They do so through the RNA sequence motifs appended to their triplex-forming nucleotides. These RNA motifs then direct outcomes by assembling proteins into a cellular machine (Figure 10.4). The exact type of machine will depend on the lncRNA motifs. A great number of machines can be generated by combining the RNA motifs attached to the triplex-forming sequence in many different ways. As these sequences do not encode critical proteins, they tolerate large variations in their makeup. There are many different ways in which new lncRNAs may form in the genome. There could be an insertion of a new sequence at the DNA locus involving the copy-and-paste mechanisms of transposable elements, by recombination, or through sloppy DNA repair. The transcripts generated then have novel combinations of protein binding motifs that can guide the assembly of novel complexes with new functions. Those lncRNAs that work best can be determined by natural selection.

Much evidence has been produced to suggest that triplex-forming lncRNAs help in the assembly of complexes that regulate the readout of genes, often in a tissue-specific and developmental way. Many of those lncRNAs conserved between human and mouse are thought to perform essential functions. The prediction then becomes that ablating these lncRNAs will change phenotype. Frustratingly, that often does not seem to happen. When there is a change, the effects vary with the strain of mouse studied. That suggests there may be redundancy, with other lncRNAs able to substitute for the absent one. Many lncRNAs may also share motifs, allowing one lncRNA to partially compensate for the loss of another, depending on how related they are with the deleted lncRNA and how the various motifs it contains are spliced together. That redundancy is not entirely unexpected given the duplication and scattering of various motifs throughout the genome. There may be a multitude of other ways to assemble the required proteins into a complex. For example, the proteins required may have a different domain that allows then to bind to an unrelated lncRNA that also forms a triplex at the location in question.

Whether there are triplex-specific proteins has also been extensively investigated. Many helicases that disassemble triplexes have been identified. Currently, there are no crystal structures of a protein bound to a triplex to provide further guidance as to the way a triplex-specific protein interacts with its target. Sequence-specific recognition of the T-flipon may only be through the RNA that anchors the lncRNA protein complexes lncRNA with no need for a sequence- or structure-specific protein to bind the targeted triplex.

L-FLIPONS

A different flipon class with an alternative two-stranded DNA structure is also of interest. It is formed by sequences that are easily bent to form an L-shape. L-flipons are recognized by proteins with an HMG box (named after the high mobility group of proteins in which this was first discovered [157]) and have effects on the assembly of protein complexes that drive gene transcription. The change in local DNA architecture alters the relative orientation of up- and downstream sequences; the L-flipons determine whether the distal sequences are close enough for interactions to occur. Most often, the binding of HMG-box proteins brings enhancers and promoters close to each other. The bent DNA can also block the propagation of supercoils from one DNA segment to another by locking the DNA duplex in place. The accumulated supercoiling due to bending can promote Z-DNA formation or instead induce the formation of the single-stranded regions that enable formation of other flipon structures. The topology of segments either side of the bend can also vary independently of each other. L-flipons thereby allow tight control of the level of supercoiling and flipon conformation across a chromosomal region and allow each to vary with cellular state.

No flipon is an island and there is the potential for competition: formation of one structure may preclude formation of another by capturing the energy necessary to power their transition. G-flipons and Z-flipons are both enriched in promoters. Often, there are multiple instances of each flipon type (Figure 10.5). How does is the competition resolved? How does that control gene expression? Quite well, is the easy answer. An example is shown below for the MYC gene.

Mutations to c-MYC cause cancers. Originally, the gene was named for the myelocytomatosis virus in which the sequence was first found. The MYC protein stimulates the outgrowth of cells called myelocytes, an early progenitor of white blood cells, that become cancerous when infected by the virus. Then Harold Varmus and Michael Bishop realized that the virus had snatched the gene from the host at some earlier time and turned it into a cancer-causing gene (called an oncogene with *onco-* meaning tumor). Subsequently, other oncogenes were found embedded in our genomes. Those oncogenes enable cells to go solo and eventually destroy the life that supports them.

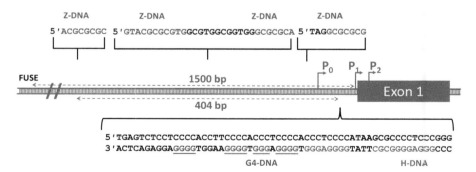

FIGURE 10.5 The human c-MYC promoter DNA is represented by the blue railway track. Regions upstream of exon 1 contain four flipon sequences that can form Z-DNA, one that can form G4Q, and yet another triplex-forming region that can fold back on itself and make H-DNA. These alternative conformations regulate another DNA segment called FUSE (far upstream element), that only binds regulatory proteins when the DNA helix opens up to become single-stranded (adapted from *Molecules*, 26,4881, 2021).

The c-MYC promoters contain the Z-flipons that Burghardt Wittig mapped with the Z22 antibody, plus a G-flipon and a H-flipon (Chapter 8, Figure 10.5). They all compete for the energy of negative supercoiling to drive the formation of alternative DNA flipon structures. As a first approximation, the ease with which the sequences flip depends on how many hydrogen bonds between the two helical strands require breaking to form a single-stranded transition state. However, once there is sufficient energy to start the flip, then competition between flipons determines the outcome. The dynamics reflect the amount of energy required by each flipon to switch conformation. Ray Kelliher and Mike Ellison examined this question by comparing two Z-DNA-forming sequences in the same supercoiled plasmid. As the negative supercoiling increased, the short Z-prone $d(GC)_7$ insert flipped first. As the negative supercoiling increased further, a longer $d(CA)_{25}$ sequence began to flip as well. Then, something interesting happened. The $d(CA)_{25}$ absorbed all the available negative supercoiling and flipped the $d(CG)_7$ back to B-DNA [158]. The switch took place even though the sequences were well separated from each other. The information was communicated by DNA from one flipon to another. It truly was action at a distance. The observation exemplified how small local changes in topology in one segment of DNA can affect DNA conformation at a site far away, changing the location at which Z-DNA forms. The change is dynamic. Why would Nature only work with static DNA conformations?

A reasonable question, but does this exchange of energy between flipons really happen inside a cell? It is a question David Levens has worked on for many years with his focus on the c-MYC gene. The key observation is that c-MYC expression is held within a tight range: too much causes cell over-proliferation and too little causes cell death. Amazingly, the c-MYC levels are maintained at similar levels in many different cell types despite differences in exposure to a wide range of environmental conditions during the various stages of cell differentiation.

The c-MYC gene has three different start sites (the promoters are labeled P in Figure 10.5). The different flipons are located within a short segment surrounding the promoters. It is unlikely that all the flipons adopt an alternative conformation at exactly the same time. The sequences are bound by different protein assemblies in different cells. The flipon conformation then depends on which sites are protein-bound as cells sense or respond to changing circumstances. How, then, is all this information integrated? The outcome depends on maintaining the negative supercoiling of the promoter within a certain range to control the rate of gene transcription. When the rate is too high, the negative supercoiling accumulated causes a set of far upstream elements (FUSE) to become single-stranded. The DNA is then bound by sequence-specific FUSE-binding proteins that prevent more c-MYC RNA from being produced. The FUSE proteins halt transcription by preventing RNA polymerase from leaving the promoter region. Of course, in a population of cells, there are those cells that express too much c-MYC and those that express too little. Neither extreme persists as those cells are outcompeted by others that express just the right amount of c-MYC and are more capable at acclimating to existing exigencies. As David would entitle one of his papers, "You Don't Muck with Myc" [159]. As we will discuss in chapter 11, flipon conformation may also be regulated by small RNAs.

11 Is Your Genome Soft-wired?

There were still more surprises in store. I thought I was done but then I heard Nagy Habid talk about the small RNA therapeutic he was developing. It activated expression of the target gene. I found that fascinating as it has long been the dream to program life directly by using only the nucleic acids from which cells are coded. For me, the question that immediately arose was whether the small RNAs were controlling gene expression by altering flipon conformation. Also, could this somehow influence the heritability of phenotypes?

It was realized early in the molecular biology era that simple regulatory schemes could match RNAs produced in one part of the genome to sequences at other locations. By binding to their target, the RNA could alter gene expression. One model proposed by Benjamin and Britten envisioned a network of RNA interactions that enabled the integration of responses to environmental events [160] (see also Chapter 15). Genetic studies in roundworms provided evidence that small RNAs (called microRNAs, miRNAs) were extremely important for regulating the stability and translation of mRNA [161]. Their discovery was unexpected, but the elegant genetic studies by the Rukvun and Ambrose labs left no better explanation for the phenotypes observed [162, 163].

The outcomes depend on Argonaute proteins that are guided by RNA to their target. These proteins are then bound by scaffolding proteins that link the complex to the various outcomes. The proteins involved are specific for the structure formed by the pairing of the miRNA to messenger RNA (mRNA). Often, the miRNA target sequences are present in many different mRNAs, providing the potential for co-regulation of their expression just by using very generic protein machinery to recognize the particular RNA structures formed between an mRNA and a cognate miRNAs (Figure 11.1).

These structures formed with an RNA produced at one site acting on an RNA arising from a different site are called *trans* interactions. The specificity is provided by the RNA, not by the generic protein effectors. Once the correct structure forms, the response kicks in, regardless of the RNA sequences involved (Figure 11.2). These *trans* RNA-directed processes are used widely in biology (Figure 11.2). The genetic code is the first recognized example. Here, the tRNA must form the correct fold with a messenger RNA codon before the cell will continue with protein synthesis. The process depends on an adaptor, first proposed by Francis Crick, that can match the triplet code of an RNA sequence to a particular amino acid (Figure 1.7). The adaptor is called transfer RNA (tRNA). The tRNA makes specific base-specific contacts with the mRNA by recognizing the nucleotide triplet (called a codon). It is charged with an amino acid at one end specific for its anti-codon triplet that specifically base

DOI: 10.1201/9781003463535-13

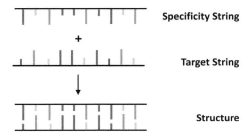

Specificity String

+

Target String

↓

Structure

FIGURE 11.1 Pairing of RNAs to produce a structure to which proteins bind without regard to sequence. The specificity is in the interaction between the two RNAs. Over time, further elaborations develop where the simple structures are encoded in different parts of the genome (adapted from *Molecules*, 26,4881, 2021).

pairs with the mRNA codon. The tRNA, once properly coupled to the mRNA, is accommodated into the ribosome at the ribosomal "A" site, allowing the transfer of the amino acid to the growing protein chain at the ribosomal "P" site. Viewed from a slightly different perspective, the tRNA binds a specific codon in mRNA to form a structure that fits into the generic ribosomal machinery (Figure 11.2). All of the tRNAs for the 20 different amino acids can form an equivalent structure when the tRNA anticodon correctly matches the codon of a mRNA. The shape formed does not depend on the tRNA sequence, nor on the attached amino acid. The correct structure enables the specific insertion of an amino acid into the proper position within a protein.

The use of a structural adaptor is a rather elegant solution to what was called the coding problem. Now this mechanism seems obvious. It was not at the time Crick proposed it, with many competing hypotheses available. The mechanism was simple in its elegance. It did not require evolving hundreds of different enzymes, each specific for joining a pair of amino acids together to make the protein, as was once favored by many biochemists (see Chapter 1). The mRNA and the tRNAs involved are made separately from each other. Yet both speak the universal genetic code and

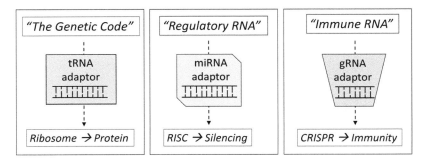

FIGURE 11.2 RNA-directed processes in translation by the ribosome, RNA interference by RNA-induced silencing complex (RISC), and the Clustered Regularly Interspaced Short Palindromic Repeats (CRISPR) RNA that directs anti-viral responses in bacteria.

both search for a perfect match. Their enduring embrace turns the ribosome on to enable the synthesis of new protein.

The same is true of microRNAs. The Argonaute AGO2 proteins will only load with a duplex of the correct length to form the RNA-induced silencing complex (RISC). They will undergo the conformational change that readies the enzyme for action only when the guide and target RNAs are aligned correctly. Then, AGO2 will make the cut to the triage the RISC-bound RNA. There are similar requirements for the PIWI RNAs that are bound by a set of proteins from a different branch of the Argonaute family [164]. The PIWI system is particularly important in restricting the spread of retroelements within the germinal tissues.

Bacteria have evolved an entirely different RNA-guided system to protect themselves against the bacterial viruses called bacteriophages (Figure 11.2) [165]. The RNAs target CRISPR-associated (Cas) nucleases to the genome of the virus to cleave DNA in some cases and RNA in others. The guide RNAs (gRNAs) are appended to the host CRISPR RNA (crRNA) sequence that pairs with another host transactivating crRNA (tacrRNA) to localize the Cas enzyme to the target [166]. Cleavage depends on recognition of a 4- to 6-base PAM (protospacer adjacent motif) in the target sequence by the Cas protein [167]. The PAM sequence must always be present and adjacent to that recognized by the guide, The correct pairing of guide and PAM sequences triggers the conformational switch that activates the nuclease. Host sequences are protected as they may contain a match to the guide, but not an adjacent PAM.

The CRISPR system is very versatile for re-engineering cells. Multiple rewrites of DNA can be performed simultaneously by coupling the cutting of DNA with insertion of particular nucleotide sequences. With these CRISPR technologies, we can now carry out wholesale multiplex changes to a genome to build an organism that is perfectly adapted to an industrial use, where enzymes replace the harsh chemistries currently deployed. Even variation to the tacrRNA:crRNA hybrid has been used to construct logic gates that allow programming of cellular responses in responses to specific inputs. The eukaryotic equivalent system called Fanzors is also guided by trans-RNA interactions (see Chapter 15).

From an evolutionary point of view, the *trans* interactions directed by RNA generate phenotypic variability rather easily. The RNA sequence space is much less restricted than the protein sequence space. You can explore the RNA space to find what works without abandoning those successful adaptations that gave an advantage in the past. The small RNAs can change phenotypes by targeting other RNAs without directly changing DNA sequence or permanently changing protein function. What is varied with microRNAs is the timing (heterochrony) and location (heterotopy) of maximal protein expression [162, 163]. These heterochronic and heterotopic alterations can greatly impact development. They affect when and where a protein acts and for how long. If the targeted RNA encodes a protein that stops cellular proliferation, a delay in expression of that protein caused by a microRNA can increase the size of an organ as more rounds of cell division are possible before the shut-off occurs.

In contrast, it is much more difficult to produce equivalent outcomes by directly changing protein sequence. Many amino acid variants will negatively affect protein function by leading to misfolding. Only a few of the variants will be beneficial. Consequently, the probability of generating desirable outcomes is much lower than those that are detrimental. The negative outcomes can be masked if one parental chromosome encodes the wild-type protein. These recessive effects will be rapidly unmasked as the variant becomes more frequent in a population. An example are the variants that produce sickle-cell anemia. When coupled with a wild-type allele, they offer protection against malaria and so have increased in frequency in regions where malaria is endemic. However, inheriting the sickle cell variant from both parents leads to devastating disease. In contrast, those maladaptive variants that dominate the wild type will survive in the population only if they are transmitted to offspring before the disease sets in. This outcome is found for multiple late-onset neurological diseases such as Huntington's disease and other repeat expansion disorders. The task of producing phenotypic variation by varying protein sequence is just so much more challenging than using RNA-directed processes to produce a much broader range of potentially beneficial outcomes.

But what made Nagy Habib's research so interesting? It was the activation of gene expression, rather than repression as seen in RNA interference [168]. The same machinery seemed to be involved, but how can you both activate and inhibit gene expression using the same pathway? Could the outcome have something to do with flipons? If so, how would that work? With Fedor Pavlov in Maria's group, we decided to start off with a simple analysis [169]. Why not see whether conserved microR-NAs (miRNAs) bound to flipons? If the interaction is important biologically, these interactions should be maintained during evolution. We started with the microRNAs that could be traced back to the era when the body plans for bilateral symmetry (as opposed to circular symmetry) arose. If there was no interaction with the conserved miRNAs, then we could rule out roles for flipons and miRNA in each other's biology. Fortunately, the datasets were publicly available and there for us to explore. Francis Collin's vision, of biology based on building the databases and they will come, was paying off for us.

We noticed a definite enrichment of matches between flipons and the seed sequences that target miRNAs. Initially, we were confused, as many of the matches were not in promoters for protein-encoding genes. What was going on, we asked. Were these matched with promoters for non-protein coding genes such as lncRNAs, or other types of unannotated promoters, or were they associated with different chromatin types? There were many possibilities! Then, we had the answer – it was the retroelements at these other locations that were targeted by microRNAs. With that clue, we found earlier work done by Glen Borchert working with Erik Larson that supported an important role for transposable elements in microRNA origins [170]. The explanation was that the microRNAs developed to interfere with the spread of these elements through the genome. Using our new tools, we had rediscovered what was found in an earlier era with less data. Our work provided an important piece of additional information. We found that the microRNAs could bind DNA flipons, something not previously appreciated.

When we looked at the genes with promoters that had flipons bound by conserved microRNAs, we were in for another big surprise. These genes were enriched in developmental pathways to a hugely significant extent [169]. So significant that a reviewer stated that, even though we used a statistical measure, the findings rose to the level of a causal relationship. We used a metric called the false discovery rate (FDR) that estimates the probability that the result we found is a mistake. As the size of a gene set increases, the measure becomes more robust because the outcome is less influenced by outliers that, for one reason or another, can bias the dataset. We saw FDRs exceeding 10^{-20} in the 3,000 or so genes we analyzed. In some cases, the FDR exceeded 10^{-100}, but only a statistician would be excited by this as that number of events is not possible in the physical universe.

Our hypothesis is that these microRNAs bind to flipons and regulate promoter shape to control gene expression. The microRNAs can either promote or prevent the flip and thereby alter the proteins assembled on the promoter. The sequence-specific binding of the RNA then directs gene expression of the developmental genes we identified.

But why would microRNAs target flipons, and why would those be enriched in developmental genes? It comes down to the question of "How does an embryo develop into a multicellular organism?". The embryo needs to bootstrap itself somehow to initiate the programs necessary to assemble itself into what it will become. An analogy is the boot program for a computer. The bootstrap code sets up the input and output of a computer so that the central processing unit (CPU) can load the operating system that enables it to execute a range of programs. Then, the CPU can take an input and process it to produce an output. In the case of the embryo, the microRNAs inherited from parents can target flipons in promoters to mark them for later use for programming tissue development.

Development would proceed following the scheme described in Figure 11.3. Initially, the fertilized egg would undergo a reset to remove traces of previous

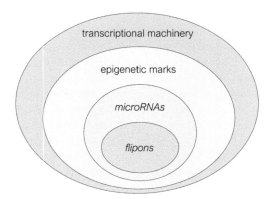

FIGURE 11.3 Bootstrap development of microRNAs and flipons. We proposed that flipon conformation is set by the small RNAs that bind to the single-stranded regions they form as they switch conformation. The small RNAs target proteins to the locus that tags the nucleosomes in the area. Later, these tags are used to guide sequence-specific transcription factors to the region to regulate gene expression in a tissue-specific manner (*Int J Mol Sci*, 24 4884, 2023).

programming. This process would then result in the widespread transcription of the genome to power the formation of alternative conformations at flipon sequences. The flips that produce single-stranded DNA would enable the suppression of endogenous retroelements by small RNAs. These non-B-DNA conformations in the promoters of protein-coding genes would also enable the binding of sequence-specific microRNAs to set the promoter state. The protein machinery assembled at these sites would depend on the class and shape of the flipons present. The proteins would then make epigenetic modifications to guide development. For example, the marks would facilitate the engagement of transcription factors that direct gene expression in a tissue-specific manner. The scheme uses RNA to kickstart the process and proteins to execute the programs at a later stage.

Of course, the microRNAs that the embryo receives from a parent play an important role in the initial development. These are loaded into each gamete and reflect the exposures an individual experiences during the period the cells are produced. They can either be expressed directly from genes active in the sperm or ovum, or loaded into the gamete as it matures through various stages. In this simple scheme, only those microRNAs that can amplify their own production in the zygote will remain at sufficiently high concentration through subsequent rounds of cell division. However, we noticed that the conserved microRNAs that bound flipons appear to be located in what is known as the extra-embryonic endoderm. These cells do not contribute directly to the embryo development (some may have a small impact), but instead act to pattern the development of the embryonic cells through their interactions with them. It seems likely that the extra-embryonic endoderm does this by producing miRNAs that are transferred to dividing embryonic cells. It appears that the basic development pathways honed over evolution are tweaked by their exchanges with the supporting extra-embryonic cast whose influence can vary from generation to generation. The design represents a different way to evolve organisms where it is not only the coding genes that count but also the transmission of small RNAs that regulate how coding genes are expressed by the embryo during development.

SHAPING DNA WITH RNA

Are there reasons to target promoters with regulatory RNAs? Yes, there are many. The regulation of promoters by small RNAs solves many problems. One particular issue arises when two different chromosomal segments, each expressing a single gene, are fused together as part of a chromosomal repair process. This event can place a new promoter upstream of the one used for the downstream gene. Transcripts that start at this new promoter will continue through any downstream promoter and prevent its use in a process called transcriptional interference. The previous regulation of the downstream gene is then lost. Small RNAs that target the upstream promoter can avoid this problem by silencing the flipons that drive its transcription (Figure 11.4). Over time, the system will evolve so that the upstream promoter is used under some circumstances and the downstream promoter at other times. From this simple scheme, based on targeting flipons with small RNAs, quite complex regulation of gene expression becomes possible (Figure 11.4).

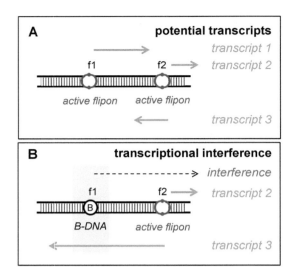

FIGURE 11.4 Transcription from flipon 2 (f2) can prevent transcription from flipon (f1), suppressing the transcription of RNA from this promoter. Targeting of f2 by a small RNA that inhibits polymerase engagement will prevent the use of f2 and allow transcription of f1. Here, active flipons are shown with a red circle containing an arrowhead. An inactive flipon is shown with B in a black circle, where "B" indicates B-DNA. (adapted from *J Biol Chem*, 299, pp. 105140, 2023).

This strategy is one that viruses also exploit. For example, the herpes viruses produce their own small RNAs. For example, the Epstein-Barr virus, the kissing virus, produces around 44 microRNAs, to redirect the cell processes to help the virus replicate, silence the anti-viral responses, and cloak its presence. Even when these viruses integrate to become a passenger in the host genome, they transcribe microRNAs to suppress their own replication, switching to a lytic phase only when they need to leave. The ability to rapidly generate new microRNAs through mutation enables the virus to evolve quickly. One strategy the cell has to respond to the disruption of the cellular microRNA function by the viral products is to suppress its own anti-viral responses with microRNAs. Any impairment to microRNA processing in normal cells by a virus will unleash an attack against the invader.

Another use of this mechanism is suggested by the presence of R-loops and Argonaute proteins at gene promoters (Figure 8.1) [125]. Many RNA transcripts fail to elongate and accumulate at promoters. Local processing of these aborted RNAs may allow their loading onto Argonaute proteins that then act to suppress further transcription from that promoter. Only when factors that enable production of full-length RNAs from the gene are present does the accumulation of aborted RNAs at promoters decrease, allowing the promoter to fully activate.

Is some organisms, nature is still capable of even more extreme strategies involving the RNA-guided readout of DNA genomes. The RNA can fix mistakes carried from generation to generation by DNA. In one well-studied case, a genome is so defective and riddled with junk elements that none of the RNAs it produces encode a functional protein. The usual response to this statement is "There is no way an organism could survive like that". But such an organism does exist. How does it pull off this impossible act? Rob Benne, Ken Stuart, and Larry Simpson explored the editing of transcripts in trypanosomes [171–173]. The editing of uridine nucleotides involved their addition to and deletion from the messenger RNA. To restore the correct reading frame for the defective RNAs produced. Of course, that process required cutting and religation of the RNA backbone to produce the corrected message. How does the machinery know where to insert or delete uridines and the number needed at a particular location? RNA guides (gRNA) were required to template the RNA repair necessary to produce functional proteins [174]. What was interesting was the cascade of sequential edits involved in fixing the defective transcript, each repair dependent on the preceding one and each requiring a different gRNA. These pan-editing events occurred in the mitochondria of *Trypanosoma brucei*, a parasite that causes sleeping sickness.

What is the sense of having a genome that is so messed up? In these situations, it does not matter. Editing of the RNA allows correction of all and any mistakes. Rather than substituting one base for another, entire pieces of the RNA are added or subtracted to rectify the problems. RNA editing corrects the code to ensure a correct outcome. Take that, you DNA supremacists -RNA makes up for your failures!

Interestingly, the editing involves two separate sets of DNA. A few dozen DNA maxicircles encode the defective mitochondrial RNAs while the more numerous DNA minicircles produce the guide RNAs. The 20–30,000 minicircles are interlocked with each other to form a kinetoplast. This arrangement ensures a high probability of transmitting all the required minicircle guides to the next generation. The set-up also allows acquisition of new guides from each mating partner, fostering the spread of new adaptations throughout the population.

Another unicellular eukaryote, *Tetrahymena*, also perform RNA-guided genome editing to render it functional. The information needed is stored in a micronucleus, which is diploid and essential for reproduction. The micronucleus produces small RNA guides, called scanRNAs during meiosis, a process that is required for mating to occur [175]. *Tetrahymena* also have a macronucleus that directs cellular functions. It forms from the micronucleus, with slicing, dicing, and selective amplification of certain chromosomal regions. Rather than using the same templates for reproduction and maintenance, the organism uses two different copies of its genome, one of which is heavily edited. Whereas the micronucleus contains five chromosomal pairs, the macronucleus can contain many hundred. Guide RNAs promote deletion of many DNA elements from the micronucleus that did not previously contribute to survival, especially those encoding retrotransposons, to make the macronucleus. The maternal macronucleus promotes elimination of any scanRNAs that have a match to the genome. The remaining scanRNAs then correspond to the DNA sequences previously deleted from the macronucleus or arising from non-host DNAs. They are then available to delete the same sequences from the new macronucleus of the next generation and to protect against viruses. Indeed, around one-third of the micronuclear genome is removed with around 12,000 DNA edits. In *Oxytricha*, another microbial eukaryote, over 95% of the micronuclear DNA is removed [176]. The process splices

DNA rather than RNA. It represents a different way to deal with your junk, allowing you to get stuff done by being more lean and efficient. Of course, RNA gives the orders and DNA falls into line.

With these examples and with the examples given as we discussed editing and splicing, it is clear that your destiny is not in your DNA. Rather, your future depends on the logic you use to read out the information contained within your genome. The wiring of your responses is not hardwired into your hereditary assortment. It is softwired and shaped by RNAs that do not code for protein.

12 How Do You Assemble a Soft-wired Genome?

It is usual to think of coding DNA as the essence of life. Therefore, our intent is to mutate exons to produce new outcomes, usually anticipating a "one and done" approach. The aim is to permanently correct a defective protein function or to engineer a genome to produce a particular phenotype. These deliberate manipulations target the codonware of a cell. The focus neglects the fliponware and wetware of a cell, the simple biology of which underlies much of what is hard to understand about how a cell functions.

Many of the reprogramming events in a cell are not hardwired nor template-driven. They are mediated through the wetware and the fliponware and involve many different forms of modifications that are not scripted directly by DNA (Figure 12.1). The outcomes are soft-wired, stochastic, and adjustable according to context. Many reponses are self-actuating and often self-amplify. The logic is implemented through the way the wetware and the fliponware assemble into scaffolds that drive outcomes. The responses are quite flexible and vary by cell, tissue, and organism.

I first visited the soft-wiring of genomes in 1999 and then again in 2004 when I discussed the importance of RNA-directed evolution. The evidence for this mechanism has grown through the work of many others we have mentioned in the previous chapters. The role of non-canonical nucleic acid structures in these processes is now quite evident. Recently, condensates, structures that construct their own compartments in cells, once considered Nobel Prize-winning, then neglected, have undergone their own renaissance. These higher-order assemblies play an important role in the chemistry and biophysics of a cell. How then do flipons and condensates contribute to soft-wired genomes?

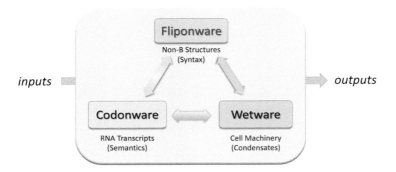

FIGURE 12.1 The cellular wares (adapted from *Trends Genet*, 36, pp. 739–750, 2020).

DOI: 10.1201/9781003463535-14

147

Let's first consider how flipons evolve over time. The simple repeats from which flipons are made are a source of genomic variability. By their nature, the simple repeats can be difficult to copy during cell division. They can grow or shrink, so each of the resulting cells receives a slightly different version of them. The importance is that the changes in length and sequence alter the ease with which they can flip to an alternative conformation. If too short, they will be forever be B-DNA, and if too long, they may freeze in a non-B-DNA conformation. In the latter case, the repeats may disrupt normal processes like transcription and replication by creating a barrier that prevents enzymes from completing their tasks. In cases where such problems arise, there are repair proteins to discipline the errant flipon. These enzymes will try to fix and eliminate the problem, either directly by trimming the sequence or by excising the segment from the genome. The processes in turn produce additional genomic variability.

In some cases, a broken chromosome will fuse to some other chromosome as the lesion undergoes repair. The resulting genomic rearrangement then changes the expression of genes in the neighborhood where the fusion took place. In the clinic, we often see the negative effects of chromosomal DNA exchange when a promoter from a tissue-specific gene is fused to an oncogene, causing misexpression and predisposition to cancer. Such transpositions can also increase a species' chance of survival. There is evidence that a fusion event gave rise to human chromosome 2, accounting for the 46 chromosomes in humans compared with the 48 chromosomes of chimpanzees. In that process, new genes arose and the expression of the existing ones changed. It is also likely that the imbalance in chromosome number created a reproductive barrier that enabled humans to evolve as a new species since the offspring produced by interbreeding were either no longer viable or were infertile. Frozen flipons may promote the chromosomal breaks that underlie such events.

A different, more targeted approach to increase genomic variability has been suggested by Lynn Caporale, who proposed that there are genomic regions selected for their mutability. They are retained because those segments enabled key adaptations in the past. The high mutation rates of repeats, some of which are also flipons involved in regulating gene expression, certainly fit this picture well. One dramatic example of such a region was found by Kathleen Xie and David Kingsley and involved the loss of a d(TG) repeat in the promoter of stickleback fish that leads to the loss of pelvic hindfins. The mutation was found in many independent fish samples taken from lakes isolated from each other. The promoter deletion is selected against in ocean fish, but is more frequent in fish from lakes where there are no predators to defend against.

The insertion and deletion of retroelements into the genome can play a more nuanced role. These events alter how genes are expressed and how RNAs are spliced and translated. The retroelements often carry along with them RNA sequences that are transcribed from their original location. They then paste this information along with themselves into their new genomic home. The flipons and other sequence motifs can subsequently alter the mix of RNAs read out from these new neighborhoods. They can change the transcription, editing, splicing, and stability of the transcripts produced from the region of insertion. The changes are not all-or-nothing. The novel messages generated can coexist with the older isoforms that encode previously successful adaptations. A larger transcript space is now available to explore. Most of the new transcripts will be intronic or defective and junked without ever being used. The increased transcript variability can foster new adaptations that facilitate the rapid updating of genetic programs. The best of the novel isoforms produced will undergo natural selection. The inherent programmability of the fliponware involved will enable faster adjustments to change than allowed for by other forms of sequence variation.

PEPTIDE PATCHES

The wetware of the cell is also directly affected by the spread of repeats. Diseases caused by repeat expansion exemplify this process. In some families, the onset of severe pathology is triggered when the number of copies of a repeat expands past a certain length. More than 40 diseases, primarily of the nervous system. are caused in this manner. These include dementias, movement, and muscle wasting disorders with names like Huntington's disease, Friedreich's ataxia, frontotemporal dementia/amyotrophic lateral sclerosis, fragile X syndrome, and Unverricht-Lundborg myoclonic epilepsy. The repeat expansion can result in abnormal DNA and RNA conformations, sequestration of RNA-binding factors, precipitation of protein aggregates, and membrane pore formation. RNAs are often transcribed from both strands of the repeat region. They can be translated in all six possible reading frames, often with one producing a toxic product that damages the cell.

These diseases raise the question of what role peptide repeats play, if any, in normal cells (Figure 12.2). Earlier studies had revealed how unstructured peptides can assemble into membrane-less condensate structures. Many have highly ordered structures. They are named descriptively. Examples include nuclear speckles, PML (promonocytic leukemia) bodies, Lewy bodies, the nucleolus, P-bodies, and many more. Often, the condensates localize various factors that process RNAs. Surprisingly, many condensates do not have well-characterized functions despite years of intensive investigation.

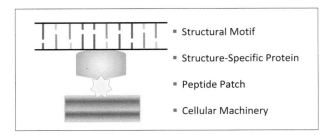

FIGURE 12.2 Simple structures formed by nucleic acids are recognized by structure-specific proteins that nucleate the assembly of cellular machines to perform specific functions (adapted from *Molecules*, 26,4881, 2021).

Recent work has focused on the seeding of condensate formation by intrinsically disordered regions of proteins. Here, we will focus on the role such peptide patches play in diversifying the wetware of a cell. Like flipons, these peptides arise from DNA repeats that spread throughout the genome, although not necessarily the same repeats that encode flipons. They alter cellular function in many different ways and are subject to selection.

Simple peptide repeats can patch together proteins with highly evolved functions to create cellular machines that perform very novel roles in a cell. Like other repeats that evolve rapidly, the length, composition, and location of patches within proteins vary. The patches do not alter the protein's catalytic rate or its preferred substrate. In some cases, they will block access of a reactant to an enzyme's active site, performing a regulatory role. Mostly, the patches are add-ons that impact the interaction of a protein with other molecules, patches can bring proteins together to perform a specific function. By their closeness, proteins can hand off a product to the next protein in the production line, increasing the rate of the overall reaction. The complexes can function efficiently even when the concentration of components or reactants in the cell is low. In contrast, some patches can hold proteins apart to prevent their aggregation.

Patches also interface the functional core of the complex with the cellular environment. The location of these tags on the exterior surface of proteins allows for their modification in a flexible fashion. There is no need for a genomic template to guide the reactions. Indeed, there are a remarkable number of ways to derivatize patches to reflect a change in a cell's internal state. The frequent revision of patched interactions allows the adjustment in real-time of responses to environmental perturbations.

Many of the peptide patch adducts seem baffling at first glance and raise questions like "Who ordered that?" Some modifications may be random as they are not precisely targeted and have no overall effect. However, a fatty acid modification to a patch may be a direct measure of lipid concentration inside the cell, allowing assessment of the current metabolic state. These processes may involve addition of butyryl, crotonyl, and other groups that provide a read-out from different pathways. The adducts can also directly alter protein function. For example, negatively charged acetyl groups can modify the positively charged lysine and arginine amino acids on histones, diminishing their binding to the negatively charged backbone of DNA. This alteration then impacts gene expression. Other modifications, like the conjugation of small proteins, such as ubiquitin and SUMO (Small Ubiquitin-like Modifier), or of carbohydrates or phosphates, also modify the stability of proteins.

The variation in amino acid composition and length of peptide patches also impacts their function. For example, patches that contain repeats of the same amino acids allow for sloppy regulation. Often, there is a cascade of modifications triggered by the initial one, with only the final adduct critical to the outcome. The early modifications often do not precisely target a particular residue. Instead, modifying any amino acids in a peptide repeat may be sufficient to trigger the subsequent steps. For example, with the ubiquitin system, phosphorylation of one residue or another may be enough to dock the required ubiquitinylating enzyme that specifically modifies a particular amino acid residue. The multiple potential phosphorylation sites that can be targeted by multiple different kinases ensure that the system is robust. The design allows the addition of ubiquitin to a particular site to occur in both a context- and tissue-specific manner, depending on how and when each kinase is expressed.

Outcomes may depend on the number of modifications made and whether or not they are all of the same kind. With single modifications, they may enable or disable an interaction. Alternatively, they can act as accumulators to sum up the total number of positive and negative votes cast for a particular outcome. This final count may in turn determine whether or not an assembly forms or a particular response occurs. The patches provide a means of converting a collection of analog inputs into a binary output.

The binary nature of the output can trigger the formation of large scaffolds that coordinate different responses through the modifications that are directed by the peptide motif involved (Figure 12.3A). The scaffolds provides a framework for the assembly of complex biological machines made with many different effectors. In the case of ZBP1, fusion of the small protein ubiquitin near the amino-terminus opens up the protein to expose the RHIM domain. The RHIM domain can then interact with other RHIM domain proteins, like RIPK1, RIPK3, and TRIF, to initiate the addition of long ubiquitin chains to RIPK1. Those chains provide a framework on which to build signalosomes. These complexes promote cell survival through the activation of mitogen-activated protein kinases (MAPK) and NFκB transcription factor-dependent expression of genes that promote protective inflammatory responses. Other proteins can disassemble these ubiquitin chains, allowing a different set of ubiquitin modifications that promote degradation of the signalosomes by the proteasome (Figure 12.3A). Alternatively, the adducts may be incorporated into autophagosomes, a system that is designed to encapsulate protein complexes within lipid membranes (phagosomes are named because they "eat" the cellular contents) (Figure 12.3B). The vesicles formed may fuse with others that promote the destruction of their contents, or instead undergo export either to the extracellular space or to other cells. The system can remove entire signalosomes from the cytoplasm to rapidly terminate their action, or expel aggregated proteins that represent a danger to the cell, thereby avoiding further damage.

In other contexts, ZBP1, RIPK1, and RIPK3 proteins can interact through their RHIMs and assemble into multimers, a process facilitated during times of cellular stress by heat-shock proteins. The fibers formed then allow the assembly and activation of proteins like the caspase 8 protease that initiates cell death by apoptosis. Alternatively, RIPK3 scaffolds may instigate inflammatory cell death by phosphorylating MLKL protein to induce pore formation in cellular membranes. The pores lead to the collapse of the salt gradient across the membrane, causing activation of inflammasomes that promote the caspase 1-mediated cleavage of interleukin-1 and -8 precursors. Caspase I also cleaves gasdermin to create pores in the cell membranes that release the processed IL-1 and IL-8 from the cell to drive the inflammatory responses of other immune cells (Figure 12.3C).

How can ZBP1 affect such different outcomes? One answer is through the concentration of the protein in a cell. Normally, ZBP1 is expressed at a low level and then only in cells of the immune system and epithelium. Ubiquitylation of the ZBP1

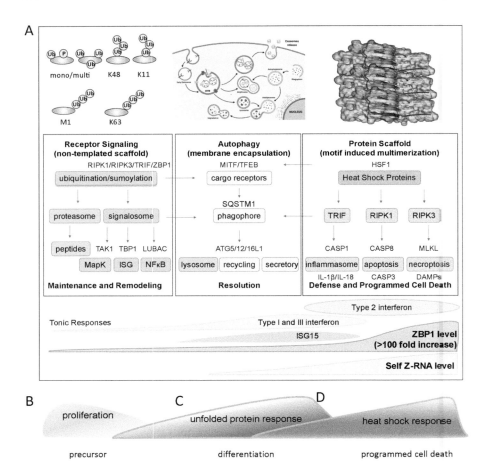

FIGURE 12.3 Different ways to scaffold outcomes using peptide motifs to interface with specific cellular pathways. A. (left panels) Adducting small proteins like ubiquitin or SUMO to a peptide motif in a protein creates add-on scaffolds to guide different outcomes. For example, K11 and K48 chains promote degradation of the adducted protein, M1 provides a framework for NFκB activation while K63 targets proteins to lysosomes. Addition of ubiquitin to peptide motifs can be triggered or prevented by other modifications such as phosphorylation. A multitude of other non-templated modifications, based on alkylation and glycosylation, also exist (middle panels). Cargo receptors can bind to peptide motifs before or after modification and initiate phagophore formation that then encapsulate proteins into membrane-bound vesicles for further processing, secretion, recycling, or destruction (image adapted from *Front Cell Dev Biol*, 12, 8, 614178, 2021, CC by 4.0) (right panels). Heat-shock proteins can promote the assembly of TRIF, RIPK1, and PIPK3 (Protein Database Structures 7DA4 and 8IB0) into higher-order protein scaffolds that regulate inflammation or cell death by activating downstream effectors. CASP1 proteolyzes gasdermin to form pores in cell membranes, while RIPK3 phosphorylates MLKL to form a larger-sized pore. Caspase 8 causes apoptosis by triggering the release of cytochrome c from mitochondria. B. Proliferation and self-renewal of precursors. C. Differentiation of precursors into mature effectors. D. Programmed cell death of infected, dysfunctional, and senescent cells. The events of B., C. and D, mirror the scaffolds in use at a particular stage of a cell's life cycle.

amino terminus is sufficient to open up the ZBP1 RHIM domains to initiate the signaling pathways shown in Figure 12.3A, B. In this context, ZBP1 promotes cell survival and protective inflammatory responses. The low ZBP1 protein level ensures that the different cellular pathways remain in balance and that only well adapted clones proliferate.

High levels of ZBP1 expression is induced during viral infection by type I interferons: ZBP1 RNA levels can increase 100-fold. In these situations, an infected cell poses a great danger to the host. Interferon also elevates the expression of retroelements that contain Z-boxes bound by ADAR1 and ZBP1. Activation of ZBP1 by self or pathogen occurs when ADAR1 is unable to suppress the increased levels of Z-RNA. The response is further amplified by the type 2 interferon produced by immune cells that further enhances RIPK3 expression. ZBP1 activation of RIPK3 then initiates RIPK3 autophosphorylation, leading to the formation of RIPK3 scaffolds that promote the quiet termination of distressed cells via apoptosis or by inflammatory cell death to terminate a viral infection (see Chapter 7, Figure 12.3C). The context therefore plays a key role in determining whether or not a cell survives and, if not, the manner in which it dies. In these situations, the Z-RNA eliminates threats by promoting cell death while in normal cells, Z-DNA plays quite a different role in transcription (Chapter 8).

PEPTIDE PATCHES AND EVOLUTION

The patching together of proteins to expand the range of possible outcomes differs from the way a well-trained engineer would design a machine to achieve the same end. Precise engineering with secure connections and guaranteed system performance are preferred by us all. Our most marvelous machines, such as a Ferrari, are designed just to go fast, but elegantly so. We start with a clean sheet and create parts that fit perfectly. We can make advances by elaborating the previous battle-tested design. Yet, that strength is also our weakness. The Ferrari is purpose-built. It has limited uses, it cannot repair itself, nor reproduce itself, nor improvise anything. Those attributes are superfluous to its purpose. In comparison, Nature works with wrecks, adding and subtracting bits and pieces over time, never stopping and always on the move. Though, in principle, Nature could trans-mutate a wreck into a Ferrari, that outcome is unlikely to be useful. The benefit of working the wreck is its adaptability, with the peccadillos from the past and the peculiarities of the present are all, somehow, put to good use in the new assembly. The simpler the pieces, the better the outcome, as such components will often satisfy sufficiently. The repurposed junk enables many different constructions, just as any assortment of stones enables a myriad of unique habitats. Unlike a well-machined Ferrari, the solution needs only be fit for purpose, not perfect and certainly not machined with sub-micron tolerances.

Despite these differences in design philosophy, the use of standard parts is an underlying feature common to how we and Nature leverage the past. They provide a framework to build from. Often, the innovations are simple, such as the manufacture of screws with a specified thread that varies little, regardless of source. Bolts designed to work with a generic wrench are another instance of a generalized solution, as are gear wheels. Once commoditized, these items enable many alternatives to build structures with off-the-shelf availability. Items can be added on, swapped out, or upgraded to create or improve outcomes. The modularity enables construction of quite complex machines from simple components of known quality. As the catalog of possible parts grows, each with its own strengths and weaknesses, so does the diversity of outcomes. Peptide repeats and their different non-templated modifications enable these elaborations in biological systems.

In Nature, simple structures enable complex outcomes. Often, the next step in the compilation depends on the structure formed in the previous step. At each stage, the outcome is determined by how tightly the components bind to each other. This process begins randomly. The accidental collisions enable sampling of the elements locally available. In some cases, the contact is fleeting, not quite right, so the partners drift apart. The off-rates in these cases are usually quite fast. In other cases, the initial touch progresses to a full embrace that is of high affinity. The slower off-rates favor complex formation. The patchwork then arises by matching each new part to those already present, creating a unique assembly from what is available.

A classic example of this process is provided by the enzyme RNA polymerase 2 that transcribes most of the protein-coding genes. The enzyme has a flexible tail made up of many heptad peptide repeats (Serine-Proline-Threonine-Serine-Proline-Serine) that are essential for its activity. The number of heptad repeats varies from 26 in yeast to 52 in vertebrates. Of the seven amino acids, five can be modified by phosphorylation. The other two residues, both prolines, can exist in one or two conformations (in *cis* or *trans* orientations). The repeats and the various modifications provide a flexible interface with which a wide range of proteins involved in transcription and RNA processing can bind to produce a variety of different outcomes. As the heptad tail changes, so do the parts cobbled together and so does their output.

The heptad repeats allow an RNA polymerase to carry with it the components necessary to process transcripts in real time. Indeed, it is thought by Patrick Cramer that quality control mechanisms involving a group of proteins called the integrator complex ensure that the polymerase is properly loaded as it leaves on its journey. If not, transcription terminates, likely leading to R-loops in the promoter, as shown in Figure 8.1. The heptad repeat may also serve a different role during transcript elongation, providing a way to reload proteins onto the RNA polymerase as each splice is initiated.

Peptide patches like the heptad repeats of RNA polymerase 2 and the negatively charged patches of histone chaperones help seed the assembly of cellular machines. The initial contact is merely the chance to go further. Patches of low complexity can interact with other repeat patterns. For example, the positively charged histone tails can recognize the negatively charged phosphate backbone of a DNA structure, or self-assemble to form a mesh, like Shuguang Zhang's zuotein peptides. That initial assembly is only the start. Concentric layers can then form, with each subsequent shell built from different patch sequences present on each coating of protein. The inner layers may be soft and pliable while the outer layers are hard and protective, walling off the space to shield what is happening inside from the outside. This arrangement has two important consequences. One is that the inner compartment can perform specific functions more efficiently than is possible otherwise, because the space is sequestered from external events. Differences in free water, pH, and electromagnetic gradients may enable chemistry not possible in the water phase. Intermediates that might disrupt the cell if not walled off are also confined. The arrangement is similar to a fireplace that stops a house from going up in flames every time a fire is lit. This compartmentalization is important for many of the metabolic reactions critical to cell function, such as those that allow the mitochondria to produce energy. When these spaces break down, cell death is often the only way to efficiently limit the damage.

Sometimes, a phase separation just sequesters proteins and RNAs away from other parts of the cell. In this case, there is a balance between what the condensate can do versus what they prevent from happening elsewhere in the cell. Stress granules are one example that traps ribosomal components and RNA messages to limit their loss when the cell senses a threat. The same process may shut down viral replication factories. Lastly, the condensates may just prevent charged surfaces from causally clumping together and forming disease-causing aggregates, like those found in the repeat expansion disorders described above.

The condensates initiated by patches may also protect the cell from the genomic instability associated with flipon transitions. Due to their single-stranded nature, the junctions between the different DNA conformations and B-DNA are susceptible to damage. The junctions can break, be cut by enzymes, expose bases to oxidative stress, or lead to adduct formation by chemical mutagens. None of these changes are good for the cell. The coating with low-complexity peptide patches can protect junctions from those assaults..

There are many examples where peptide repeats and fliponware collaborate to self-organize higher-order structures. This cooperation is evident during the assembly of ribosomes, machines that translate RNA sequences into proteins. The process is extremely complicated and involves putting together a large and small subunit, each composed of numerous proteins. The subunits form on their own ribosomal RNA scaffolds, each of which must also be correctly folded for the process to work. The assembly occurs in the nucleolus, a nuclear structure without membranes that is seen as a blue spot when the nuclear DNA is stained with a fluorescent dye called DAPI. The nucleolus consists of three layers, each of distinct composition. One layer performs DNA transcription, another mRNA processing, and the third assembles ribosomes. In the first layer, T-flipons engage the lncRNA named PAPAS to regulate ribosomal gene transcription. In the next layer, the ribosomal RNAs undergo base modification that is guided by small nucleolar RNAs that recognize specific sequences. The complex formed incorporates proteins, like fibrillarin, which is a methyl transferase. Fibrillarin modifies the ribosomal RNA sugar to reduce the capacity of the RNA to induce interferon responses triggered by the dsRNA regions formed as the ribosomal RNA folds. The outer nucleolar layer can be visualized with fluorescent dyes that stain G4Q and is organized by proteins like nucleolin and nucleoplasmin that recognize and resolve G-flipons to correctly place additional ribosomal proteins. Each layer has its role to play in the production of ribosomes.

Surprisingly, Alu repeat elements also are enriched in the nucleolus. Indeed, the formation of the nucleolus appears dependent on this set of Alu sequences because knocking their level down disrupts the structure. The Alus required derive from introns transcribed during the production of mRNA. The mechanism may match ribosome production to the levels of gene expression: higher Alu levels lead to the assembly of more ribosomes. The Alu elements may also contribute structural elements, although this is not proven. The Alus contain short G-sequence elements that can hydrogen bond with similar sequences in other Alu fragments to form mini quadruplexes that can stack onto each other when aligned to form G4Q wires. The A-rich linker between the two Alu monomers can also form triplexes.

Although the ribosomal genes contain an abundance of Z-flipons, there is no ribosomal editing by ADAR1 and no activation of ZBP1 in normal cells. The nucleolar structure prevents the editing enzyme and their potential Z-DNA targets from finding each other. The way nucleoli are organized also appears to prevent the induction of cell death by Z-DNA- or Z-RNA-induced necroptosis in normal cells. Only when the nucleolus undergoes disruption during stress or viral infections are Z-flipons unmasked. The induction of cell death then limits viral replication and propagation.

Spliceosomes are condensates defined by a combination of shape and sequence. They are composed of an ensemble of heterogeneous ribonucleoproteins (hnRNP) and include essential splicing factors that incorporate small RNAs (called snRNPs, the acronym for small nuclear ribonucleoproteins) used to guide assembly of the spliceosome. In this process, each snRNP binds to a defined element of the pre-RNA. The correct alignment of RNAs within the spliceosomes is required to catalyze the cleavage and joining together of the splice junctions. The spliceosome also engages proteins that perform base modifications and RNA editing. The process is sequential, allowing completion of one processing step before the next. For example, the editing of a dsRNA editing substrate formed from an intron and an exon must occur before splicing. It is likely that flipons help in timing these events by directing the p150 isoform of ADAR1 to this subset of pre-mRNA substrates, especially when Alu inverted repeats direct the base pairing between intron and exon sequences.

A subset of hnRNPs in the spliceosome bind to simple sequence repeats in the RNA through RNA Recognition Motifs (RRMs). The hnRNPs also have low-complexity positively charged patch residues composed of serine, arginine, and glycine that initially localize the protein to highly structured RNAs such as G4Q RNAs. The hnRNPs then help dock the cellular machinery necessary to unwrap these RNA folds to enable binding of the hnRNP RRMs to the single-stranded RNA produced. Assembly of the complex occurs in a highly cooperative fashion to promote formation of the splicing scaffold. The G4Q structures appear to be especially important for the splicing of short exons.

The alignment of splice sites can involve the looping of chromatin from distant chromosomal regions, bringing these sequences close together. The loops involved are often a megabase or more in size and are formed within topologically associated domains (TADs). The challenge is to align the correct set of splice sites. There are many other potential donor and acceptor splice sites in the intervening region that could be used, but are not. The sliding of one chromosomal segment past another is one way to produce the required alignment of different sites, but what causes the process to stop and perform a particular splice? Although G4Q may contribute to this process by localizing the helicases involved in assembling the spliceosome, other flipon types could also be important for correctly aligning splice sites. For example, the bending at junctions where Z-DNA abuts B-DNA may help bring into close proximity proteins bound to distant regions of the gene. . Alternatively, a change in flipon conformation may point the DNA arms away from each other and diminish interactions between the splice sites. Evidence for a triplex-based mechanism that can potentially align sites within TADs was uncovered by Gary Felsenfeld, who discovered the triple helix at the dawn of the molecular biology era. In this new work, he was investigating the *in vivo* regulation of hemoglobin gene expression. He found that an intronic RNA reached back to form a triplex within the promoter region of the gene. The interaction down-regulated expression of the transcript. Others have also produced evidence consistent with a role for long noncoding RNAs (lncRNA) in gene regulation. The lncRNAs also appear to bridge two distant sites through a triplex formation. Similar interactions could also help approximate splice sites. Methods to track these connections are rapidly improving.

RNA contributes to nuclear structure in other ways. The integrity of the nucleus itself depends on RNA. Jeff Nickerson and Sheldon Penman found long ago that nuclear structure is disrupted by RNase treatment. When the RNA is removed, everything else in the nucleus clumps together. In contrast, removing DNA with DNase does not lead to matrix disruption. Repetitive RNA elements are likely involved as they can form stable hybrids with each other. They can assemble into higher-ordered mats for peptide repeats to anchor the chromosomal loops while different segments slide past one another. The RNA matrix provides the stage upon which the genes "are merely players". (*As You Like It*, W. Shakespeare)

Overall, these processes are based on structures encoded by repeat sequences. These elements provide versatility and adaptability in the fliponware and wetware responses to environmental changes. The peptide repeats undergo a variety of modifications to seed a plethora of protein assemblies. The different assemblies allow cells to optimize responses during periods of stress by altering flipon conformation. The cells that survive may not necessarily be the best available, but are just good enough to endure the existing exigencies. Given a different challenge, a different set of cells may prevail.

Through interactions mediated by peptide patches and alternative flipon conformations, a large set of possible response repertoires are rendered. Just as it is impossible to hardwire sufficient antibody diversity into the genome to survive all current and future microbial and viral threats, it is also impossible to hardwire the best possible responses to every environmental challenge. The soft-wiring of cells ensures that no two cells are ever the same. It allows selection of those cells that are best adapted.

13 How Do You Program a Soft-wired Genome?

The possibility of using biological organisms as computational devices has intrigued many now that bioengineering has become a routine activity. Many of the approaches are based on the lift-over of the Boolean AND, OR, and NOT gates similar to those used in *in silico* devices to implement logic circuits (Figure 13.1). From these designs, it is possible in principle to construct all other logic gates. Those logic gates built from nucleic acids variations come under the rubric of string operations. The strings are the letters for the sequence of bases in an RNA or DNA molecule. Manipulating strings can involve adding, subtracting, or rewriting one or more of the bases. These processes can be guided by hybridization of one nucleic acid strand with another. The guides specify the logical operation performed at each step. These reactions rely on Watson-Crick base pairings between the two or more DNA or RNA strands coupled with processes to ligate, cut, or edit the strings.

Quite complex programs can be implemented just using DNA and hybridization of one strand to another when each sequence has base complementarity with the other. The solution to the traveling salesman problem is an example that attracted much interest when implemented by Leonard Adleman in 1994 [177]. The challenge is to find the shortest, non-overlapping path between multiple cities without ever visiting a particular city more than once. The combinatorial possibilities make this problem difficult to solve computationally. It is possible to use DNA to find a solution as the search can be performed massively in parallel. The answer is provided by using hybridization to search all possible DNA sequences to find the one that is shortest. This process involves starting with a string for each city that also includes a representation of each road leading into or out of a city, then hybridizing the city strings to a DNA pool containing all possible solutions. An enzyme is then used

Flipon Computational Gates
Logical Operations and Strings
– AND
– OR
– NOT
Topological Switches
– Action at a distance
State Switches
– On/Off
– Toggles

FIGURE 13.1 Logical operations for building DNA computers.

DOI: 10.1201/9781003463535-15

to ligate the annealed city strings and the routes connecting them. Only those city strings bound immediately next to and connected by a single path between them will be joined. The answer is provided by the shortest ligated DNA fragment that contains just one connection between each city string. The result is read out by sequencing the DNA. The limitations of this approach include the sheer bulk of DNA involved in the calculation, the time necessary for hybridization reactions to occur, and the concomitant increase in error rate in the hybridization reaction as the number of cities visited grows larger. Overall, these DNA computers do not scale well in time or space with an increase in the number of possible combinations.

Other designs conditionally switch "on" or "off" expression of genes to provide a binary output. The regulation can involve a small molecule that controls the production, stability, or editing of an RNA or protein in combination with a guide RNA required to target the desired modification to DNA, RNA, or a protein scaffold. Sensors can be linked to reporter genes that provide readouts for each particular event. Multiple AND gates are used to reduce the error rate and increase the specificity of detection, while NOT gates can prevent false positives. A number of examples of biocircuits based on these principles have been published by the Domitilla Del Vecchio, Ron Weiss, and Jim Collins laboratories at MIT. These designs elaborate on the approach Norbert Wiener called cybernetics [13]. Here, a causal process is controlled by feedback loops where the output moderates the processing of inputs but is not itself an input to the circuit. Negative feedback circuits limit the output according to how much product accumulates, while positive feedback loops amplify outcomes.

Other approaches based on RNA editing and splicing events can also be employed in logical circuits to produce a conditional RNA output. Logical AND and OR gates schemes can be implemented using RNA splicing to perform the operation. In Figure 13.2, the NOT gate uses an exon that encodes a toxic product when read out as a protein. If not removed, the cell dies. The AND gate joins two particular

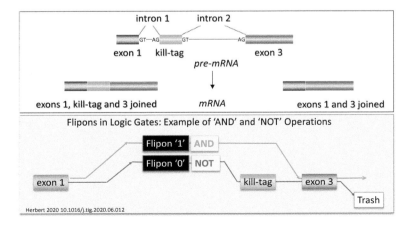

FIGURE 13.2 Flipon "AND" and "NOT" gates based on splicing (adapted from *Trends Genet*, 36, pp. 739–750, 2020).

exons, separated by the toxic exon that is excluded by the splicing event. The OR gate would represent any splicing event that excludes the toxic exon but does not require the joining of any particular pair of exons. The approach provides a large space for innovation. Tools are available to moderate the selection of splice sites and a number of protein products provide a suitable readout after splicing is completed.

Of course, logic gates controlled by splicing are found in Nature. One outcome is the determination of sex in flies (Figure 13.3). Specification of a female depends on the correct splicing of the double-sex protein that depends on the correct splicing of the transformer 2 protein that depends on the correct splicing of the sex-lethal protein. Ultimately, the sex depends on them having an equal number of X and non-X chromosomes. It is the inclusion of exon 4 in double-sex that is the key event. Sex-lethal and transformer 2 both catalyze their own splicing to create a positive feedback loop that drives the pathway to produce a female. The RNAs are produced by both sexes but the splicing is what makes the difference [178]. The logic is soft-wired and each step is conditional on the processing event that went before.

RNA editing can create splice donors (e.g., AT →GT) or remove splice acceptors (AG → GG) to change the splicing a pre-mRNA undergoes (Figure 13.2), as can blocking of an acceptor or donor site with a protein or a small base-complementary RNA. Another possibility is the use of small molecules to change the way an RNA folds. This can affect the availability of splice sites and determine whether the RNA is translated or degraded. A number of riboswitches, turned "on" or "off", by small molecules have been characterized by Ron Breaker at Yale [179]. More complex RNA structures , called ribozymes, can act as enzymes, and perform the steps

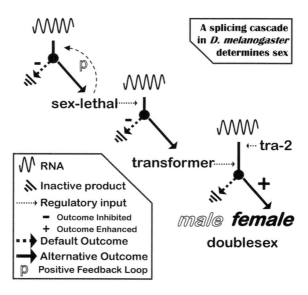

FIGURE 13.3 Determination of fly sex by alternative splicing. The female sex depends on the proper splicing of transformer pre-mRNA by double-sex protein and its proper splicing (from *Ann N Y Acad Sci*, 18, pp. 119–32, 1999).

required for cleavage, replication, or ligation of RNA [180, 181]. The activity of these ribozymes is dependent on forming the correct RNA fold and can also be modulated with small molecules. This approach offers a wide range of options for implementing soft logic with RNA strings.

There are additional possibilities enabled by RNA editing that affect the translation of RNA into a protein used as the output of a circuit. For example, the expression of a protein can be controlled by editing a start codon. In this case, the AUG that initiates translation can be edited to GUG, preventing expression of a reporter protein. The result can be easily quantitated using a downstream, non-edited AUG that directs the expression of a different protein. The presence of an upstream AUG in the RNA transcript prevents use of the downstream AUG: the alternative protein is not produced. If the upstream AUG is edited, translation of the protein from the downstream AUG then occurs. The amount of protein synthesized from the downstream AUG then allows quantitation of how much editing enzyme is present in a cell. I used this approach in 2002 to make a case that some RNAs were translated in the nucleus [182]. It is now being developed for cell-based sensors where the level of editing enzyme expression depends on an environmental input.

Other ways of reprograming cells by RNA editing include the recoding of specific residues in proteins. A protein function, or its interactions with other molecules, can be altered with a single amino acid substitution. The amino acid recoded could be essential for the enzymatic function of a protein. In cases where a Mendelian disease is due to a variant with diminished activity, then editing of the RNA to recode the residue offers a precise way to correct the error. For some proteins, modification by phosphorylation or another adduct is essential for an interaction with their targets. Recoding the residue to prevent this modification would prevent the assembly of a complex that enables a particular response. The editing event allows these processes to be halted temporarily as the RNA will eventually be turned over and replaced with new, unedited transcripts. The intervention can then be timed to make edits that turn off pathways that exacerbate a pathology.

A different programming approach uses flipons as topological switches to change the output of a mechanical device [183]. For example, the flip from A-RNA to Z-RNA alters the length of a double-stranded RNA helix from 24.6 Å to 45.6 Å [3]. The flip thus changes the distance between the two ends of an RNA that contains an embedded flipon. Since the flip is rather rapid, changes can be detected in the millisecond range. If the transition is induced by stretching the DNA, then the flipons can act as nanoscale strain sensors. An RNA containing multiple flipons of equal length could also act as a cumulative strain gauge, enabling fine-scale measurements in real time. The effects of thermal noise and other stochastic variations in flipon conformation are a potential problem at this scale of magnitude, but can be overcome by a statistical approach to find the average conformation of a number of individual sensors.

Flipons may also find use as state switches to amplify a signal through the pathway that they turn on or off. In chapter 6, I described how the flip to Z-RNA can turn off the interferon response. We also discussed how flipons can alter the expression of genes. In these responses, the synthesis of both RNA and proteins is involved, and the response time will be minutes to hours. When the flipon acts as a toggle switch to cause a state change, the signal will be amplified as the number of switched cells increases as cells divide. The design may then be extremely sensitive and enable both the detection of small amounts of ligands and the scanning of large volumes for their presence. The different types of flipons may each be suited to different applications. Quadruplexes, for example, are quite stable due to the extensive hydrogen bonding involved. They offer potential as memory elements.

Programming of flipon-dependent switches with small RNAs is also a possibility [169]. These RNAs bind to their single-stranded intermediates formed as flipons transition from one conformation to another (see chapter 11). With G-flipons, an RNA binding to one strand may free the other to fold into a quadruplex. The G-flipon may promote the assembly of a complex that promotes the expression of a gene. One example is the binding of the ICP4 protein to quadruplexes formed in the herpes virus genome. The interaction tells the cellular machinery where to begin transcription of RNA. Quadruplexes also localize repair complexes to damaged DNA. In contrast, binding to the G4-quaruplex-forming strand by a small RNA will prevent quadruplex formation. The location where a G4 quadruplex forms can also change outcomes. When present in a gene body, the folded structure can act as a barrier to a transcribing RNA polymerase, which envelopes both strands of the DNA and has its passage blocked at the region of G4Q formation. The use of RNAs to modulate the G4Q conformation can allow programming of these outcomes. Flipons represent a new approach to the design of devices for biocomputing.

Engineered circuits based on DNA and RNA logic gates have many potential applications. Already under development are biocircuits for the detection of pathogenic bacteria. These engineered approaches detect proteins or toxins produced by the pathogens, Signals are enhanced by the use of enzymes that self-amplify and sensor cells that replicate rapidly. Both the specificity and sensitivity of detection can be enhanced further by screening libraries of genetically encoded biocircuits to find those performing best. A wide variety of reporter proteins are available for these applications. Biocircuits to sense a therapeutic response to a drug or to control delivery of a therapeutic are also approaching the clinic. In one example, bioengineered cells can release insulin when blood glucose levels rise, but then stop delivery once the glucose concentrations fall below a particular level. While these controls could be engineered into patient cells, the biocircuits could be delivered as cell-packs of non-self cells encased in a membrane that minimizes the risk of an immune response against their contents. Such devices, once implemented, are cheap to make. The cells can be fabricated on a large scale in a bioreactor, reducing the cost of manufacture and delivery. Furthermore, the devices are inherently self-repairing using the fixit pathways that Nature has evolved over the eons. The pack may be implemented as a wearable device. Alternatively, the pack design could enable subcutaneous implantation, with rapid removal and replacement if necessary. Local delivery of therapeutics to the alimentary track by orally delivered packs may be a different option for gastrointestinal diseases.

The approaches so far described mirror our current computational designs. The logic is transitive, based on A leads to B, then to C. The schemes modify the cell operating system, but there are limits on the numbers of such circuits that can be built into a single cell. However, there are more intransitive ways to go!

14 Is Life Intransitive?

The reason to examine intransitive logic in cells is that this approach leads us to a different view of how life evolves. Rather than focusing on how pathways change through one mutation or another, or on how duplication or loss of a particular gene modifies expression of a trait, the emphasis is on how cells regenerate their components. Attention is then paid to directed cycles (DCs) that couple unrelated chemistries in ways that generate new functionalities. The logic involved is intransitive with examples found in a wide range of natural systems. The DCs and their intransitive logic greatly impact the evolution of cells.

So, what is intransitive logic? This form of logic is familiar to all who play the game of rock, scissors, paper. In this contest, there is not a single strategy that always wins. The outcome is determined by what the other player choses. For example, rock beats scissors, scissors beat paper, but paper beats rock. The only way you can always win is if you know what the other player will choose, then you can beat their move with yours.

Transitive logic can be stated as A>B and B>C implies that A >C (> indicates greater than, or an event that occurs before the other). If the rock, scissors, paper game used transitive logic, A would always beat B and C, and B would always beat C. Therefore, you could guarantee a win by always playing A. With intransitive logic, the relationships are expressed as A>B>C>A with C> A. where C beats A even though B beats C and A beats B. The intransitive relationship between A, B and C is illustrated by the directed cycle drawn in Figure 14.1 This DC only flows counter wise (the convention we will use in this book), as indicated by the arrow head. When the rock, scissors, paper game is played with intransitive logic, C in fact beats A, despite A's dominance over B and B's dominance over C. Your expectation of a win is one time in three if both players choose simultaneously. However, if the other player chooses first, you can always find the option that wins the game for you. Stated differently, there are some responses that will win in one situation and lose at other times.

So, how would intransitive systems work in biology? Before I answer the question, let me provide a background to the problem. I will start with a brief historical introduction to these ideas and how they apply to DCs. Then, I will focus on DCs that remain in a stable state, but never reach a point of chemical equilibrium. Naturally, we need to answer the question: "How is this energetically possible?". If a system is running downhill, why does it not hit rock bottom? We can then ask the question of how DCs are used by cells for computation. Of course, you could skip directly to the section starting, "So how can DC be used computationally?", but then you would miss out on finding an answer to a mystery of the ages: "Why does life exist?" and "How does life persist?"

DCs depend on the continuity of intransitive logic. You can enter at any point and leave at any other point, but there is no beginning or end. Of course, directed cycles

DOI: 10.1201/9781003463535-16

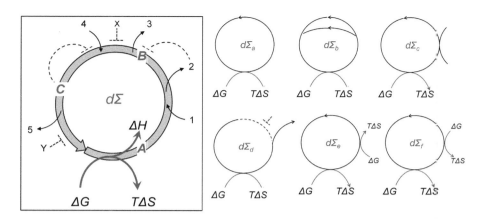

FIGURE 14.1 A directed cycle that implements intransitive logic where you can start and leave from multiple points. They capture the relationship A>B>C>A. There is no beginning nor end to the cycle. The letters A, B, and C could represent the rock, scissors, rock response. The cycle depends on energy input (ΔG). It maximizes work (ΔH) by minimizing entropy loss (TΔS). The dotted lines represent a subset of possible paths that allow negative regulation of the cycle through elements B and C, or through points X and Y. In Nature, these cycles are quite stable and can be described as a class of dissipative structures (dΣ). Different ways to utilize dΣ are labeled with a subscript: a, reference; b, redundancy; c, connected; d, interrupted; e, downhill; f, uphill. If the cycle looks messy, then you understand the point being made about how biological systems evolve (from Int J Mol Sci . 2023 Nov 18;24(22):16482).

are not perpetual motion machines. They are not like Penrose's impossible staircase (illustrated well by M. C. Escher) where the stairs sit atop four walls set at right angles to each other. The design allows you to either always go up or to always go down, much as the artists draw them. The direction you go depends on your choice and, since there is no change in height during each cycle and the system is ideal, no energy is used, meaning you could continue forever. DCs are different. They go only in one direction, but they require an energy input to return to their starting point; otherwise, they stop and do not regenerate their components.

Where does that energy to power DCs come from? The following discussion will be loosely based on the work of two Nobel Prize winners: Ilya Prigogine and his concept of the dissipative structures, and Manfred Eigen and his work on hypercycles. The focus is on biological systems that are far from chemical equilibrium and that regenerate themselves.

Prigogine coined the term "dissipative structures" (dΣ) to describe self-organizing systems that dissipate energy to maintain their stability [184]. The dΣ represent states that have low entropy relative to their surroundings and remain highly structured despite the widely fluctuating inputs they receive from their disordered surroundings. As they are far from chemical equilibrium, there is sufficient free energy available to dΣ to offset the entropic cost of retaining their ordered state. In many cases, dΣ are able to switch from one stable state to another, with small perturbations often sufficient to trigger the transition.

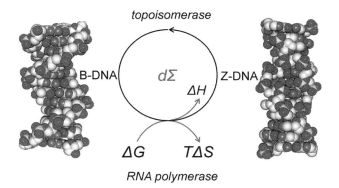

FIGURE 14.2 Flipons as dissipative structures. They represent directed cycles between right-handed B-DNA and left-handed Z-DNA conformations. Polymerases can provide the energy to initiate the flip from B-DNA to Z-DNA. Dissipation of energy by topoisomerases relaxes the Z-DNA to the B-DNA conformation (from Int J Mol Sci . 2023 Nov 18;24(22):16482).

Dissipative structures come in many forms and DCs are but one of the possibilities. Flipons are one example that require energy to drive the transition from B-DNA to Z-DNA (Figure 14.2). The energy can be generated through the action of RNA polymerases. The polymerase produces mechanical energy as it unwinds and negatively supercoils the DNA helix, while releasing chemical energy by breaking phosphate bonds as it incorporates nucleotides into RNA. The energy is captured by the flip from B-DNA to Z-DNA and is released as the Z-DNA flips back to B-DNA. The energy stored in Z-DNA can power completely unrelated events, the nature of which depends upon the path taken as Z-DNA relaxes back to B-DNA. For example, the flip to B-DNA can fuel a change in chromatin state, enhancing or inhibiting transcription. Alternatively, the energy can be dissipated by topoisomerases, ensuring that flipons do not freeze in the Z-DNA conformation, thereby reducing the risk of strand breakage at the B–Z DNA junction.

More complex dissipative structures can form in completely different ways. Many involve quite complicated chemical pathways. One famous example is the Belousov-Zhabotinsky (B-Z) chemical reaction. Rather surprisingly, the reaction mix repeatedly changes color from red to blue and then from blue to red and so on as the solution is mixed by constant stirring. When first discovered, the pattern of color oscillation was quite unexpected. The manuscript describing the reaction was widely rejected. The journal editors were certain that the reaction scheme violated the second law of thermodynamics. The second law states that for any spontaneous process, the total entropy of a closed system either increases or remains constant; therefore, one of the two states in the B-Z reaction, but not both, had to have a lower entropy than the other. However, this is not true. The observed oscillations are no different from those a swing undergoes. These die down as the kinetic energy is expended by frictional forces. Consequently, to maintain the height of the initial trajectory, it is necessary to supply energy by pushing or pumping the swing. Note also that the B-Z

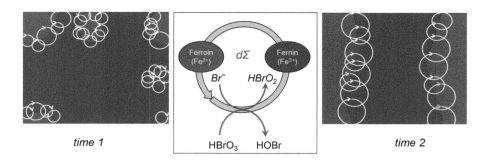

time 1   time 2

FIGURE 14.3 A plot of phase changes. Computer modeling of the Belousov-Zhabotinsky reaction shown in the center panel. The red and blue color changes correspond to the oxidation states of iron (Fe). The left and right panels show snapshots of the system at *time 1* and *time 2*. The boundaries mark the region of transition between the two phases and are very sharp. The B-Z cycles that define the boundaries are drawn in white. The conversions that drive the reaction are given by the red arrows. The reduction of iron from blue to red is driven by the autocatalyticformation of $HBrO_2$ in the solution (from Int J Mol Sci . 2023 Nov 18;24(22):16482).

reaction depends on the autocatalytic production of $HBrO_2$ rather than as a reaction product (Figure 14.3).

During the B-Z reaction, a sharp transition occurs between the two-color states. This process can be modeled using deterministic equations that underlie the transition from one state to the other. Figure 14.3 contains two snapshots from a computer simulation of the reaction. All the chemical components are free to diffuse within the solution as there is no physical separation between the phases. The figure illustrates the color variation over time at different positions in the reaction space. At a critical concentration of the bromide ion, the reactants initiate a step-like switch from one phase to the other. This occurrence is found in many systems. For example, the patterning of biological organisms, such as seen with zebra stripes and in angel fish, was proposed by Alan Turing [185] to also arise from chemical gradients that produce sharp boundaries between phases. The gradients were later and independently confirmed by Hans Meinhardt and Alfred Gierer in 1972 [186]. The pattern arises from the interaction between a slow-diffusing autocatalytic activator of the reaction and a fast-diffusing inhibitor.

Another famous demonstration of DCs was given by Robert May in 1976 [187]. He built on previous work that examined the effects of population doublings coupled with the loss of breeding partners along the way. The change in population size is described by a deceptively simple equation. The number of individuals at the next time point is given by:

$$x_{t+1} = r x_t (1 - x_t)$$

where x_t is the proportion of the maximum possible population at time "t" and "r" allows for different rates of population increase. The term $(1-x_t)$ accounts for those

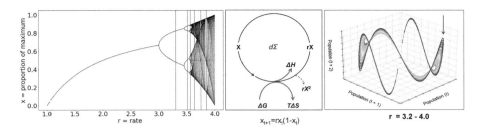

FIGURE 14.4 Population recurrence map. A. Swings of maximum and minimum population size at a replacement rate r. B. At each step, the population grows by rX and decreases by rX^2, where r is the rate of increase and X is the ratio of the current population to the maximal possible number. C. The cycles for r = 3.2 – 4.0 for 3,000 cycles measure at times t, $t+1$, and $t+2$ (from Int J Mol Sci . 2023 Nov 18;24(22):16482).

individuals unable to breed due to death, starvation, pestilence, or war. The equation can be plotted out for different values of "r", showing the variation in maximum and minimum population size. As shown in Figure 14.4, there are values for "r" where the population oscillates between stable maxima and minima (as indicated by the vertical red lines). At other times, the maxima and minima population sizes show no such regularity. Again, the regions of stability appear within very choppy seas. Yet, these widely different outcomes are described by that simple and innocent-looking equation.

An idea of the stable oscillations in population size can be obtained by looking at a 3D plot of 3,000 paths as r varies between 3.2 and 4 where the maximum and minimum population size is relatively constant. The plot follows the oscillations in population number from t through $t+1$ to $t+2$ (Figure 14.4), mapping the path followed during each individual cycle. The ups and down between relatively fixed points are quite evident. The coloring indicates that the paths followed are non-overlapping yet the peaks and valleys are very similar over time. The vertical arrow points to a region of the plot where it is easy to see the separation between paths. In this region, the increase in entropy with each cycle is quite apparent. Here we measure entropy as the number of states that are available to the system. At the scale drawn, we can mostly resolve each path. However, with time, as the number of paths become greater, it becomes harder to go back and retrace the exact history. Even if we could continue to magnify the image, we eventually reach the Planck limit where we cannot precisely define the different path histories. As the entropy increases, we can only follow the system forward in time but we cannot retrace its past. As we cannot reverse the timeline, the system conforms with the second law of entropy.

Similar patterns of oscillations between regions of relative stability were later described by Edward Lorenz in weather simulations performed by computer [188]. He called the stable regions in his graphs "attractors", noting that a small change in input would shift the system from one stable region to another. The effects were dramatic, with the system behaving chaotically. In these simulations, a small change of input could have either a small or a large effect on the outcome, depending on whether or not the input triggered a transition between states. Yet, the effects of each input were completely described by fully deterministic equations, not by random effects.

While the oscillation between stable states is characteristic of chaotic systems, it does not mean that the time spent in each state is equal. By varying the energy barrier that separates states, one state can be favored over the other. A way to imagine this is to consider two attractors, one at the top of a hill and the other at the bottom. Since going uphill is harder than going downhill, the system is biased. The slope and height of the hill will determine the amount of work necessary to move from the lower basin to the higher basin. As this cost increases, more time is spent in the lower basin.

It is worth noting that non-chaotic systems also exist that oscillate between stable states. These systems have very low entropy, i.e., have fewer paths to follow. They are very sensitive to fluctuations in input values. Consequently, they are not robust and break easily, with small fluxes pushing the system beyond the bounds of stability. The paths then rapidly diverge from each other. Periodic inputs help protect non-chaotic systems against complete chaos by confining the cycle paths to a narrow course. The tapping of a spinning top to maintain its vertical alignment provides an analogy.

Ilya Prigogine's work concerned the energy flux through dissipative structures. In Figure 14.1, the input of energy (ΔG) is explicitly shown. The ΔH represents the work necessary to complete the cycle, whereas the term called entropy (ΔS, a measure of the system disorder produced by the cycle) represents the energy lost to the environment at the particular temperature (T in degrees Kelvin) studied. Of course, this leads to the equation of Willard Gibbs that succinctly summarizes the energy balance:

$$\Delta G = \Delta H - T\Delta S$$

For work to be done during the cycle (i.e., ΔH is positive), then ΔS must also be positive. In other words, the increase in order of the system is exchanged for disorder of the environment. Of course, a source of the free energy G must be available to make up for losses if the cycle is to continue running. We have already seen that an increase in entropy over time is inevitable from cycling alone. The energy lost must be replenished from somewhere to maintain the cycle.

There is clearly plenty of energy available to systems that operate far from equilibrium. Under those conditions, life is possible. Conversely, attaining thermodynamic equilibrium with their surroundings is fatal for any living organism. There is nothing more final than the transformation of all your available free ΔG into $T\Delta S$. Life depends on dΣ that minimize entropy loss and maximize the work performed. They evolve these structures over time to improve their chances of survival. Those cells that fail do so soon fade into the void.

To remain viable, living organisms must regenerate all their components. They are prone to break as losses of key elements are unavoidable. They are also tasked to produce materials consumed by other processes. They must balance their outputs with the inputs they receive.

dΣ based on DCs allow cells to avoid the infinite regression that Bob Rosen noted in 1959, where, to make a component, you require an enzyme; to make that enzyme you need another enzyme; and to make that enzyme, you need another enzyme, etc. DCs are quite flexible and solve for stoichiometry in a variety of ways. They can receive input and produce output of components from any part of the cycle (Figure 14.1). There are many opportunities to procure parts that they cannot replace themselves. In some cases, a downstream input will eventually regenerate a missing upstream input as the cycle reiterates. The input could also be sourced from the environment, from another cell, the output of another DC, or other reactions (Figure 14.1, dΣ $_c$). Conversely, limiting the availability of an input provides a strategy for regulating a DC output.

DCs can capture energy at multiple steps (Figure 14.1, dΣ $_e$). They can drive steps in the cycle that are thermodynamically unfavorable by ensuring products from the DC are kept at low concentrations, pulling the reaction forward (Figure 14.1, dΣ $_f$). One example involves channeling a product through a membrane so that it is not available to drive the reverse reaction. The production of proton gradients across mitochondrial membranes is based on this strategy. The flow of protons in the reverse direction is then coupled with ATP production. Of course, the chemiosmotic theory formulated by Peter Mitchell to explain these events was rejected by "strong characters with weak arguments", as was noted by Wolfgang Junge [189]. DCs can also incorporate cybernetic controls, including negative loops as shown by the dotted lines in Figure 14.1. Such refinements maintain the cycle in balance so that regeneration occurs at every possible turn.

So how can DC be used computationally? We can start with Boolean logic by assigning "0" and "1" to the absence or presence of an input or an output. This assumption is reasonable for enzymes that respond in a step-like manner when a certain substrate concentration is exceeded. The directed cycles then can be viewed as a series of logic gates through which AND, OR, and NOT functions are implemented. In this case, the transitive relationships between input and output nodes can be used to construct a truth table. It is also possible to build conditional relationships using DCs. For example, in Figure 14.1, the output from an input at 4 can be 3 or 5, depending on whether or not an inhibitory signal at X or Y is present. Depending on how long 3 or 5 remain high, the DC can provide a short-term memory of exposure to X or Y, resembling how some neural circuits respond to stimuli (Figure 14.1, dΣ $_d$).

What makes this system different from a computer-based solely on transitive logic? For a DC, each relationship between an input and an output is only a subset of the logical operations that the cycle can perform. With a purely transitive design, the relationship of input to output is fixed. In contrast, the intransitive logic of a DC allows a node to assume many different roles. The node can be both an input to the DC, an input to the next step in the DC, an output from the preceding node in the DC, or an output from the DC.

Hence, the truth table derived from an intransitive system depends on the roles assigned to each node in a DC. Although the wiring is set, the order of information processing by a DC is not. The upstream node defines the path to the other node. The steps taken are different when the roles for each node are reversed. The design allows the DCs to access different cellular resources along each route and to do different types of work along each path. In that process, many different outputs from the DCs can be generated and the energy expended can be minimized over time through evolution. The DCs efficiently do what is needed for a cell to survive at that given moment in that particular context.

Engineers have tried using computers to model DCs in biological systems. One question asked is: "Can you evolve computers to make them better?". The initial approaches were based only on transitive logic. Many different designs have been tried. One implements a set of competing programs to perform a particular task. A metric is then used to find the subset of programs with the best performance. These programs are then bred together to produce progeny programs that then undergo the same selection process. Mutations to and cross-overs of the code are performed to model what happens to the DNA within chromosomes. For example, bits may be flipped or code segments exchanged. There are obvious limitations to this approach: as Andrei Kolmogorov proved, no string representation of a program can yield itself if the complexity of the code after evolution is less than the complexity of the starting program. How many bits are needed to evolve a better version of a program is also not knowable at the outset of a project as there is no sure way to guarantee that particular outcome. Consequently, you will never know whether a system of the complexity you hope to evolve is possible with the resources available to you. Nor will you know when you have found the best possible solution as you can always generate another program that may be better but that you have not yet tested.

More recent computational approaches use a massive set of connected nodes to implement systems that have transitive properties. The systems are tuned using a range of machine learning approaches to optimize a particular outcome by using a cost function based on their output to optimize their input by using a cost function to minimize errors in their outputs. Training minimizes both the energy loss and the entropy cost to maximize performance. In this sense, these deep learning approaches model $d\Sigma$, but only transitively. As such, they are stable to a range of inputs, but they can break down just as $d\Sigma$ do when certain inputs produce unexpected outputs. Currently, these systems are limited when compared with those in living systems. The DCs used by cells are self-powering, self-regenerating, self-repairing and self-referential. They model what is happening outside the cell by the changes they produce inside the cell. Their outputs allow a cell to be self-aware and self-responsive. The intransitive nature of DCs enables cells to behave in ways that are beyond the capabilities of our currently manufactured computational devices.

The intransitive logic of cells renders them programmable and evolvable in a different way than is possible with transitive logic. The differences arise because DCs are inherently self-referential. Because of this characteristic, many true, but apparently contradictory, logical schemes can be drawn to map a DC component to itself or to other outcomes (Figure 14.5). There are many ways possible to depict the relationships. A component can map to itself ($d\Sigma_1$) or to another component that is not itself (($d\Sigma_2$); likewise, mapping of a different component can be to any other component ($d\Sigma_3$) or to itself ($d\Sigma_2$). Although these mappings are all true, it is possible to take pairwise combinations of the relationships that are on the surface contradictory, leading to those vexing existential questions such as "Does x cause f(x) or does f(x) cause x?"

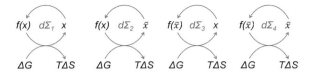

FIGURE 14.5 Different logical representations of DCs. (from Int J Mol Sci . 2023 Nov 18;24(22):16482).

Gödel noted similar issues with whether self-referential statements are provable when encoded by a formal mathematical systems based on Peano arithmetic (see https://plato .stanford.edu/entries/goedel-incompleteness/). While this is a problem for computers that are expected to stop once they reach the end of their program, living systems are not designed to halt. The DCs can represent both true and false statements, depending on the context. However, as noted for other $d\Sigma$, DCs do fail, not with a whimper but with chaos, with entropy ending the hollow emptiness of T.S. Eliot.

DCs can evolve over time in many ways. One strategy that DCs rely upon involves the targeted mutation of the genes encoding their components. The effects of mutation may be at the level of gene expression, on transcript processing, protein turnover, or protein modification. Over time, protein interactions and enzymatic activities undergo adaption to ensure survival. Recently, these refinements are often the result of bioengineering rather than evolution.

Experimental approaches aimed at modifying DCs depend on first identifying the elements essential for the DC operation. The studies can be performed *in vitro* by purifying components and reconstituting the DCs from these parts. These approaches helped elucidate many of the DCs, such as the Krebs cycle, involved in cell metabolism. The studies can also be performed using genetic approaches to identify cycle components. Over the years, bacteria and yeast have proven particularly powerful in establishing many of the factors that modulate DCs in single cells.

Collectively, these approaches identify proteins essential for regenerating DCs at each iteration. The methods also uncover redundancies and scaffolds that enhance the performance and robustness of DCs (Figure 14.1, $d\Sigma_b$). Furthermore, the results inform which DC steps can be modulated therapeutically. Drugs to break DCs are part of the pharmacopeia positioned to kill cancer cells. The targeting approach yields valuable insights into the differences between normal and diseased cells. This work identifies redundancies in normal tissue that are no longer present in cancer cells. Mutations that inactivate one or more of the redundant pathways make tumors vulnerable. The tumors are then susceptible to drugs that target the remaining pathway. Collectively, the drugs and mutations synergize to selectively kill the tumor. Normal cells survive drug treatment because they retain both redundant pathways.

Drugs that induce synthetic lethality in tumors are important in the clinic. In many cases, tumors are able to mutate and become resistant to most drugs that are used as a single agent. The tumors then continue growing. A drug cocktail that targets multiple DCs to induce synthetic lethality through different pathways is often needed to thwart the escape of cancer cells from eradication. The challenges of curing cancers despite the high-precision targeting of molecules underscores the overall resilience of cells. The intransitive programming based on DCs enhances their adaptability. Winning strategies just require a rewiring of the path between two nodes. In some cases, cancer cells even run DCs like the Kreb's cycle in reverse by wiring the nodes into a DC that promotes their proliferation. Metabolites for their growth are prioritized over an efficient supply of energy. They gain power by reprogramming normal cells to do the work, exporting miRNAs though nanotubules and small vesicles to ensure the nutrients they require are produced in abundance.

There are many highly diverse strategies to evolve directed cycles from scratch. The peptide patches we have discussed as part of the cell's wetware (Chapter 12) can act as Velcro at early stages of evolution, creating new assemblies by pulling proteins together. Initially, they may self-assemble in various ways, some potentially catalyzing the formation of peptide bonds to further their own production by acting as templates for further peptide assemblies. These structures are likely the origin of the many self-assembling filaments that enable a cell's movements and provide the scaffolds that organize cellular responses (Figure 12.3). They can also cause amyloid depositions in disease. Like their modern counterparts, the scaffolds formed also brought different sets of proteins together. The outputs from one of the sequestered proteins then potentially acted as input to another one. Eventually, a self-sustaining cycle arose through protein interactions that positively reinforced each other's output.

This developmental timeline assumes that proteins are more multifunctional than is currently presented in textbooks. In reality, the patched-together proteins often contain multiple different domains. Though many domains have well-studied functions, others remain uncharacterized. With the patchwork design just described, peptides with no enzymatic function are able to create new opportunities. The interactions unmask proteins with multiple personalities, enticing them to reveal a different character. Frequently, experimentalists find surprising the newly discovered behavior of a previously well-characterized protein. They then write papers entitled "Hidden protein functions and what they may teach us" [190] and "Protein moonlighting: what is it, and why is it important?" [191].

The new cycles established by patching proteins together may initially depend on inputs from the milieux to bridge any missing links. The Krebs cycle that we depend upon to extract energy from sugars likely developed in such a way. At an early stage, the reactions depended on environmentally derived metals for catalysis. More efficient reactions arose when binding sites for metals were incorporated into genetically encoded proteins. Of course, there are DCs that not only regenerate a component but also output that component for use by other DCs. One example, first noted by Tibor Gánti in 2003 [192], is the glyoxylate cycle, in which malate uses the energy available from acetyl-CoA to both regenerate and output itself from the DC.

$$\text{malate} + 2\text{acetyl-}S\text{-CoA} + 3H_2O \rightarrow 2\text{malate} + 2H\text{-}S\text{-CoA} + [6H]$$

This design favors the evolution of a different DC that uses malate as an input (Figure 14.6).

Today, many of our essential nutrients reflect our need for those factors. The dependency is so complete that, without them, certain DCs fail to regenerate.

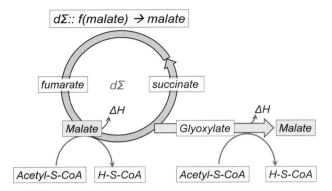

FIGURE 14.6 The glyoxylate cycle regenerates and outputs malate with acetyl-*S*-CoA pulling both steps. (from Int J Mol Sci . 2023 Nov 18;24(22):16482).

Humans, for example, do not synthesize vitamin C due to a gene mutation. They rely on other organisms to supply their needs.. Through such strategies, Nature can build new DCs by exploiting the excesses of their environment. The process can increase robustness by creating alternative links in the cycle to bypass any blocks that arise (Figure 14.1, dΣ $_b$). The synthesis also allows other DCs to evolve that depend on a particular output from an existing DC (Figure 14.1, dΣ $_c$). The input can then be generated in other ways. The new strategy allows a cell to carry on as usual. At the same time, the cell can uncover new chemistries that open up new opportunities to exploit.

We do not know how far the patchwork approach can be pushed to engineer new DCs. First, can we expose existing DCs to alternative chemistries to create completely new reaction schemes that have never before existed in Nature? Already, DCs have been adapted to use synthetic chemicals in preference to their natural substrates. For example, Madeleine Bouzon and Philippe Marlière evolved one particular metabolic pathway to use 4-hydroxy-2-oxobutanoic acid as the carbon source rather than substituted serine or glycine. We have no idea what Nature can do when put to the test [193]! Secondly, can we randomly tag proteins to generate new protein assemblies and evolve DCs with a desired output (Figure 14.7)?

These approaches elaborate on proposals made in the past by other scientists. Manfred Eigen focused on the organization of self-replicating molecules connected in a cyclic, autocatalytic manner [194]. Due to the way they interact, the cycles become self-propagating, with each cycle in a node coupled into a larger cycle (Figure 14.1, dΣ $_c$). The interactions between different cycles allowed them to amplify themselves, each other, and the hypercycle (Figure 14.7, left panel). The hypercycles further favor systems that store the information necessary to continuously regenerate themselves. In the simplest form, the earliest steps in a pathway did all that was required to undertake the next step. Steps were added that closed the circuit, leading to self-amplification of the hyper cycle. The hypercycle underwent further elaboration by connecting to other cycles that further assured their mutual perpetuation and increased the complexity of outcomes (Figure 14.1, dΣ $_c$). The creation of genetic

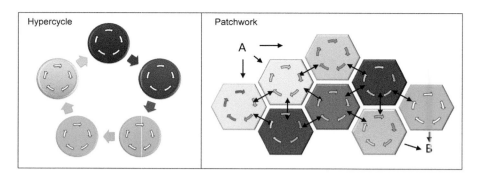

FIGURE 14.7 Interactions of directed cycles to produce hypercycles that are autocatalytic, or a more generalized patchwork that self-assembles to expand the number of connections between inputs and outputs. (from Int J Mol Sci . 2023 Nov 18;24(22):16482).

systems to transmit this information to subsequent generations was a natural consequence of this hypercycle evolution. Leslie Orgel performed studies early on to show how small molecules could autocatalyze their own production by also acting as templates [195]. Even earlier, others, like Butlerow in 1861, demonstrated autocatalytic reactions that yielded simple sugars. A 2013 survey by Andrew Bissette and Stephen Fletcher outlined many other possibilities for the autocatalytic generation of the DC s early in evolution [196].

DCs can evolve through genetic systems different from those Eigen imagined. The rewriting of directed cycles in DNA during evolution can occur in many ways. Substituting proteins with others that are better catalysts is one approach that offers a selective advantage for organisms that already know how to make both proteins. Alternatively, fortuitous mutations may help provide improved performance of the existing proteins.

Or there may be more complex processes involved. On occasions, genes may undergo duplication in ways that Susumu Ohno showed were important during evolution. Each protein replicate subsequently acquires different mutations that, at some point, can provide a selective advantage, leading to the use of one or the other in a particular DC or the creation of a new DC variant. Occasionally, whole genomes undergo duplication. Many plants have a history of expanding their genomes in this way and are consequently highly polyploid. As a result, every cell has multiple copies of each gene. The process allows DCs to be reconstituted in different ways or with different combinations to generate new elaborations. The process of genome duplication has also been observed in yeast following a sudden and adverse change in an environment. The high mutation rates that accompany this process drives additional genomic diversity and the elaboration of novel DCs that enable survival.

Another way to acquire all the components necessary to make a new DC is simply by obtaining all of them in one step from another organism. With bacteria, this means gaining an entire operon where all the genes required for regulation, expression, and scripting of a cycle are organized into one DNA segment. These outcomes are enabled by bacterial conjugation, the prokaryote version of sex first observed by Joshua Lederberg and Edward Tatum [197]. To do the same in eukaryotes would require a genomic organization similar to the operons of bacteria and a truly giant virus to transmit the much larger eukaryotic genes that embed all the required information. In fact, one of the exciting discoveries of sequencing DNA obtained from rock, soil, water, and undersea thermal vents is the discovery of giant viruses the size of bacteria that infect eukaryotic cells. It is now even possible, using a variety of technologies, to introduce into cells large genomic assemblies with all the genes required to reconstruct a DC. The most extreme transplant of genes so far performed is the transfer of an entire normal mitochondria to replace the defective ones transmitted to an embryo from a parent. Of course, the only reason eukaryotes have mitochondria in the first place is by subsuming at one point in time the whole set of DCs that another free-living organism had successfully evolved. The most recent proponent of this idea was Lynn Margulis, who also noted that chloroplasts are symbiont cyanobacteria [198]. Of course, if something is so useful, why stop after the first success? Many cells still retain the capacity to import mitochondria released from other cells. Osteoclasts that dissolve bone can rid themselves of mitochondria that are no longer functional and replace them by capturing those encapsulated within vesicles during their export from a population of circulating supply cells called osteomorphs. It is thought cardiomyocytes can also enhance their performance by taking up mitochondria from the extracellular space [199].

The genetic encoding of DCs ensures the transmission of successful adaptations to future generations. The inherent programmability of DCs enables the survival of individuals over shorter time scales. Each DC can undergo optimization as an organism finds its niche. Conceptualizing DC as a major unit in evolution focuses on the way these $d\Sigma$ enable the adaptability essential to an organism's survival. DCs trade energy for information and minimize dissipation and death due to entropic losses. Despite the perpetually fluctuating environment, DCs ensure stability by resisting change.

DCs differ from Turing machines. They are not designed to terminate. Their purpose is not to solve a problem and halt [201]. Rather, DCs work best if they never stop. As dissipative structures, DCs offer the best way to avoid a chaotic ending, but come with no guarantees. DCs embrace intransitivity and they enforce energy efficiency. DCs are not just the cycles of life but they also embed the logic of life

DCs are self-referential in that each component regenerates itself ($f(x) \rightarrow x$). Paradoxically, the junk in the genome enables such complexity. As Andrei Kolmogorov proved and as noted earlier, it is not possible to program anything more complicated than the length of the longest coding sequence available to you [200]. DNA repeats that code by changing conformation dramatically increase the complexity and programmability of genomes. They are a feature of evolving systems and nothing else. The flipons they embed enhance survival by exchanging information for energy and entropy for resilience.

15 Are RNA Therapeutics the Wave of the Future?

The 2023 Nobel prize to Katalin Karikó and Drew Weissman celebrated the milestone achievement where RNA encoded vaccines were used to blunt the SARS-CoV2 pandemic. The award recognized decades of work that contributed to the development and delivery of these new RNA medicines, advances delayed by the widespread skepticism of the initial ground-breaking studies by peer-review panels. Using RNA rather than protein has accelerated the delivery of other new vaccines and protein therapeutics to the clinic by their ease of manufacture and their low cost of delivery. There are, however, different types of RNA-based therapeutics that are now entering the clinic or have already gained approval from the Regulatory authorities. These are the advances that I will discuss here.

The story starts back in the 1960s. In the early studies on the composition of genomes, before the era of DNA sequencing, investigators would break DNA into fragments of around 450 base pairs. A salt solution containing the DNA then would be heated until the DNA duplex was melted into single-strands. As the solutions slowly cooled, a sample taken every so often would be passed over a column of hydroxyapatite to separate single-stranded from double-stranded DNA. The work allowed an estimate of how much of the genome was repetitive and what portion coded for protein. The repeat DNA, due to its high frequency, would rapidly find a complementary strand of DNA and reform the double-helix. The less frequent protein-coding sequence would not do so until much later, reflecting the time it took to find its pairing partner. With this data in hand, a model was proposed in 1959 by Britten and Davidson whereby RNAs produced by one subset of repetitive elements coordinated the tissue-specific expression of genes. Although not enough was known at the time to correctly state the details, the principle of RNA as a regulatory element of gene expression was established. Since then, the work performed over many decades by thousands of scientists has made the therapeutic programming of cells by RNA a reality. We now have RNAs in the clinic to treat a variety of diseases that target other RNAs, and even DNA, in a sequence-specific manner.

The best-known application of RNA uses the CRISPR system to directly edit DNA inside cells using a guide RNA (gRNA) to make the therapeutic DNA modifications. The approach repurposes a system bacteria use to protect themselves from viruses. The bacteria incorporate a DNA fragment of the virus into their genome within the CRISPR locus. The point of insertion is precisely determined. The viral DNA fragment is placed adjacent to a bacterial sequence that encodes a host RNA that will anchor the proteins needed to cut the viral DNA target. The bacterial RNA polymerase then copies the fusion of viral and CRISPR DNA into a single gRNA that localizes the proteins needed to eliminate the virus. The targeting depends also

176

DOI: 10.1201/9781003463535-17

This chapter has been made available under a CC-BY-NC-ND license.

on recognition of an additional nucleotide sequence in the virus DNA (the PAM sequence described in Chapter 10) by the Cas nuclease. Binding to the PAM sequence is required to activate the enzyme to cut the viral DNA identified by the gRNA. The PAM sequence is not present in the bacterial CRISPR locus, sparing the host genome from attack. The PAM sequence also restricts the targeting in the human genome and has led to much effort to find a way around this limitation. CRISPR proteins that are triggered by a variety of different PAM sequences have been engineered.

CRISPR is just the beginning as other related RNA-guided DNA nucleases have been discovered. What prokaryotes can do also turns out to be done by eukaryotes. A different class of related proteins, called Fanzors, have recently been discovered through sequence homology searches with over 3,600 unique types found in a survey of metazoans, fungi, choanoflagellates, algae, rhodophytes, unicellular eukaryotes, and viruses [202, 203]. The Fanzors are much more compact (400–700 amino acids) than CRISPRs (1000–1600 amino acids), making them much easier to package into clinically useful viral vectors [204].

The discovery of CRISPR was relatively recent and the technology is already progressing towards the clinic. However, CRISPR was not the first RNA-guided approach designed for use in the clinic. Martin Egli, who was an MIT colleague, recently reviewed the history of these approaches (Figure 15.1) [205]. Small RNAs have been approved that inhibit the translation of a cytomegalovirus protein. Others prevent splicing of RNAs in Duchenne muscular dystrophy and spinal muscular atrophy patients. These RNAs bind directly to the target. Another RNA therapeutic binds instead to the VEGF protein and inhibits blood vessel formation in age-related macular degeneration. The therapeutic is now not used as antibodies perform better in therapy due to their much higher picomolar affinity. Another therapeutic is part RNA and part DNA. The combination promotes degradation of a target by forming a DNA:RNA hybrid, a structure which is specifically attacked by an enzyme that cuts the RNA strand. The therapeutic is designed to destroy the apolipoprotein B RNA in the treatment of familial hypercholesterolemia.

The most successful applications of therapeutic RNAs exploit a different RNA pathway. The RNA also causes degradation of the targeted messenger RNAs by RNA interference (Chapter 11). These drugs are referred to as RNAi and utilize a post-translational gene silencing pathway first discovered in the nematode *Caenorhabditis elegans*. The RNA guides Argonaute proteins to the messenger RNA that is then cleaved. First developed to treat Mendelian diseases like hereditary transthyretin amyloidosis, more recent applications of RNAi reduce hypercholesterolemia in a broad range of patients by targeting the PCSK9 mRNA. The Argonaute proteins are also guided by other classes of RNAs, including microRNAs, to DNA and increase RNA transcription of the targeted gene (Chapter 11). These newer RNA therapeutics are under clinical trial. Approaches based on flipons to program cells are at an early stage.

Also new are the small RNAs designed to target ADAR1 for recoding of specific RNAs by RNA editing to ameliorate diseases arising from a single base change in the genome. The promise of this approach is that the same outcomes can be produced as in the CRISPR approach, but without rewriting the DNA sequence. The edits last as long as the recoded RNA persists in a cell. After that, the effects on the

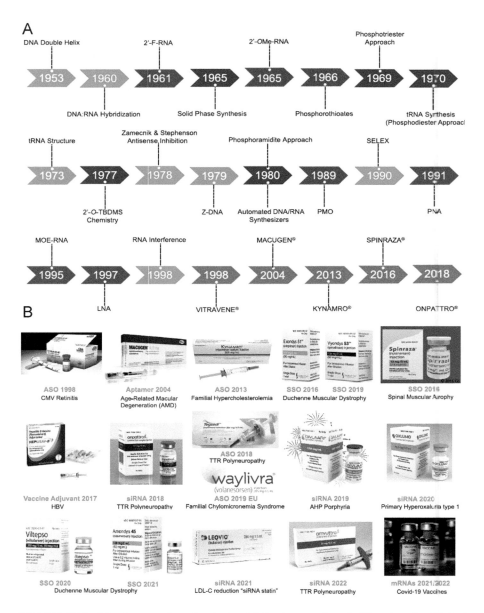

FIGURE 15.1 The long road to developing RNA therapeutics. A. Timeline. B. Clinical stage RNA therapeutics (from *Nucleic Acids Res*, 51, pp. 2529–2573, 2023).

translated protein last until the protein is replaced by a newer version made by an unedited RNA.

The first therapeutic approaches used for adenosine-to-inosine recoding were based on an engineered deaminase domain. I had shown in 2001 that the deaminase domain was sufficient for editing to occur and that a double-stranded substrate of only 12 base pairs allowed editing to occur efficiently when the RNA was expressed from a small circular DNA introduced into cells (Figure 4.10). The work showed the minimal requirements for an editing event to occur. With the bioengineered deaminase approach, the enzyme was attached to another protein domain that was known to bind tightly to a specific RNA sequence. The idea was that, if that specific RNA sequence was added to the guide RNA, then it would target the engineered deaminase to a specific mRNA. This approach was used first by Thorsten Stafforst, then the Tsukahara and Rosenthal groups [206]. The Nakagawa group added double-stranded regions from known editing substrates and attached these to the guide RNA, with the intent of using those dsRNA to recruit endogenous ADAR1, avoiding the need to introduce an additional protein construct [207]. Overlooked was the demonstration by Tod Woolf in 1995 that it was possible just to use a chemically modified, unstructured guide of 52 bases to direct editing by matching the target. Somewhat discouraging at the time was the low editing efficiency and the presence of many off-target editing events [208].

With the new chemistries available for making synthetic guides and the rush of money that poured into gene therapies spurred on by the promise of CRISPR, the talented individuals who had worked on RNAi turned to commercializing RNA editing. The majority of these RNA editing companies focused on the design of gRNAs suitable for use in the clinic. The targets were Mendelian diseases where a need existed to correct a single nucleotide variant by substitution of an inosine for an adenosine that was causal for disease. Editing of adenosine would address the unmet need in some 45% of Mendelian diseases caused by a single base change. The task was to optimize the chemistry and to eliminate off-target effects, where adenosines elsewhere in the mRNA or in other RNAs from other genes are edited unintentionally. Another approach used gene vectors to express gRNAs in cells. The advantage of this strategy is that gRNAs could be optimized by using massive parallel screens to identify guides capable of efficient editing an RNA with minimal off-target effects. The data generated then allowed the training of machine learning algorithms to predict effective guides for a wide range of potential targets.

A different approach, derived from CRISPR, repurposed the deaminase domains of RNA editing enzymes to edit DNA rather than RNA. In these designs, the cutting activity of a CRISPR system was replaced with reengineered enzymatic domains able to recode the targeted DNA base. The advantage was that editing of cytosine by deamination to thymine was possible, increasing the range of treatable diseases. Finding a way to modify cytosine bases was challenging. The enzymes that normally perform cytosine edits in the cell act as part of an anti-viral defense. They do so without RNA guides, using proteins specific for the way their substrate folds. The modified CRISPR strategy is the first RNA-guided approach able to specifically target the editing of cytosine. A subsequent approach, called prime editing, uses a reverse transcriptase enzyme to replace a segment of DNA. by copying an RNA sequence targeted to the site by a gRNA.

The length of the RNA guide required for the different adenosine-to-inosine approaches also varies. With the synthetic chemical guides, the obtainable yields of a pure, full-length, gRNA decreases as its size increases. Yet, the most recent report from Prashant Monian and Chikdu Shivalila indicate that a gRNA as small as 30 bases with appropriate chemical modifications is sufficient to produce robust editing [209]. There was no need for an added sequence to recruit ADAR1 to make editing more efficient. Off-target editing of the targeted mRNA could be eliminated by mismatching adenosines with guanosines in the gRNA. For gene-vector-delivered guides, the required gRNA length is longer, with off-target sites eliminated by careful guide design. Currently, the guide transcripts are 60–200 bases long. The stability of the expressed transcript can be increased by circularization. Here, the transcript incorporate RNA sequences called ribozymes that catalyze the ligation of the ends together. Consequently, the circular RNA lacks the free ends that are latched onto by RNA breakdown enzymes. With the CRISPR base editors, the gRNA is similar to those used for other DNA applications.

A concern with RNA editing has been the off-target edits that impact messages entirely different from the therapeutic target. Any increase in interferon associated with the introduction of the guide into the cell will naturally result in edits of self-transcripts that ADAR1 naturally makes to turn off interferon responses. An increase in such edits is not a concern and will occur at a low frequency in repetitive elements. Other edits that occur at high frequencies in messages not intentionally targeted can be detected by sequencing all the RNAs from a treated cell. If present and of concern, then a redesign of the gRNAs is likely to address the problem. In the case of synthetic guides, a change in chemical modification may also eliminate the off-target edit. A similar approach has been successfully used to assure specificity of RNAi therapeutics.

The use of DNA base editors carries the risk of introducing off-target effects not related to editing but that are permanently written into the genome. A number of sequencing strategies have been developed to efficiently find off-target sites for CRISPR cleavage events. These include *in-vitro* selection libraries (CIRCLE-Seq), chromatin immunoprecipitation (ChIP) (DISCOVER-Seq), anchored primer enrichment (GUIDE-Seq), *in-situ* detection (BLISS), translocation sequencing (LAM PCR HTGTS), and *in-vitro* genomic DNA digestion (Digenome-Seq and SITE-Seq). Cells with DNA modified outside the body by treatment with CRISPR enzymes can then be checked for problems unrelated to base editing before being put back into a patient.

Each RNA guide strategy requires different delivery approaches. The synthetic guides build on previous experience with RNAi delivery by injection. Usually, a subcutaneous route is preferred from which the RNA leaks over time and is taken up by cells as naked RNA. The RNAs are small and are also rapidly lost in the urine. Much of the work has focused on improving targeting, such as by adding a receptor-binding moiety to the guide RNA. The prime example is *N*-acetylgalactosamine that promotes uptake by the liver. Folic acid derivatives are also used. Other formulations embed the RNA in a lipid coat that bears ligands on the surface for proteins that are specifically expressed on the target cells. These ligands range from small molecules to large proteins such as antibodies, the use of which has been validated in other clinical applications. The gene vectors that deliver guides for adenosine editing and CRISPR systems also use clinically validated delivery systems, usually in a non-replicating viral vector. The viral systems can also be engineered to express tissue-specific ligands on the viral particle surface. In principle, editing can be restricted to a particular cell type by using promoter sequences to ensure that the expression of the RNA guide is tissue-specific. Overall, targeting reduces the risk of side effects and allows for much lower dosing by concentrating the therapeutic at the site where it is most needed.

Each RNA guide strategy also addresses potential adverse effects differently. The chemistries for the synthetic guides are by now well tested in the clinic. Toxic effects and immune responses against these therapeutics do not limit their use, although some patients opt out because of local reactions at the site of injection. With viral approaches, delivery is into the blood system. It is not yet clear what is the optimal dose of virus and whether immune responses to viral proteins will limit redosing.

A key question is, how long will the edits last? The experience with RNAi provides some insights. The synthetic guides are resistant to nucleases that would destroy normal RNAs. Indeed, the RNAi for PCSK9, in treatment of hypercholesterolemia, is given once every six months. The RNA accumulates in cells over time. They are mostly present in lysosomes, small bags of chemicals designed to break down used cell parts and various molecules that are internalized, along with the surface receptors on a cell. However, these enzymes do not break down the synthetic RNA guides. Instead, the RNAs leak from lysosomes to keep the Argonaute proteins loaded and ready to fire. For RNAi, it appears that, once the RNA is engaged by the Argonaute protein, it remains bound. The complex can turn over many times to cleave multiple target transcripts. For ADAR1, a similar accumulation and leakage of the RNAs occurs. What is uncertain is how rapidly the RNA guides turn over. The synthetic RNAs complex first with their target and not with ADAR1. The study by Prashant Monian and Chikdu Shivalila detected a prolonged persistence of editing in the liver of non-human primates. Editing of the target was approximately 50% at day 5 and 35% at day 45. With DNA base editors, it is also not certain whether they are a "once-and-done" therapeutic or whether repeat treatments will be necessary. A permanent fix would require the editing of stem cells from which other cells in the tissue arise. Outcomes would then be fixed in the lineage and would not revert with time.

An alternative approach to prolonging the effects of RNA-based therapeutics is to have self-amplifying RNA therapeutics. These are based on a cassette system that uses a viral RNA polymerases to specifically drive expression of the payload from a viral promoter not present in the human genome. The approach is suitable for dividing cells where the concentration of a non-replicating RNA would decrease as the cell numbers increased. A variation is also being proposed for non-dividing cells. Here, the therapeutic RNA is co-expressed with proteins necessary to package the RNA into a virus-like particle that will be secreted from the cell. The Arc protein involved in the plasticity of memory formation is one such example of an RNA delivery system that uses a retroviral capsid gene embedded long ago in the human genome and then evolved to protect neurons and enhance their function (see Chapter 7). By engineering this class of packaging proteins, sequences that target the secreted virus-like particle to a particular cell or tissue type can be incorporated, ensuring the continuous delivery of the RNA therapeutic to a specific cell type. This approach carries with it concerns about the safety of this strategy as the amplification of oncogene encoding RNAs incorporated into the cassette during the normal cycles of DNA damage and repair is a possibility. Regulation of gene expression by orally available small molecules is one approach for mitigating this risk.

Other strategies are under development to turn off RNA-directed editing when problems arise. With chemically modified RNAs, a second RNA that binds and inhibits the gRNA can be administered. The gene vector therapies can also be modulated by small molecules that turn RNA expression on and off as required to optimize the level of editing required. The CRISPR-based approach can mark the surface of cells with a second edit that allows for the elimination of modified cells. One form of this strategy that is under development involves the recoding of hemoglobin genes responsible for sickle-cell anemia. In addition to editing the DNA base that corrects the disease, a second edit is performed on the CD117 cell surface receptor gene that is expressed only by hemopoietic stem cells [210]. The edit does not affect receptor function, but instead allows for these cells to be specifically targeted by an antibody that does not recognize unmodified cells. A second antibody exists that has the opposite binding properties; it does not bind the recoded receptor but recognizes only the unedited one. The second antibody allows for depletion of the disease-causing stem cells without having to kill them using toxic chemicals prior to the infusion of the modified cells. The first antibody allows tracking of what is happening to the edited stem cells once they are delivered to the patient. If necessary, the first antibody can be used to eliminate the edited cells and allow their replacement with cells lacking the recoded receptor. The approach promises to make bone marrow transplantation less hazardous and also lower the cost of the procedure. The strategy potentially provides a revolutionary change in the practice of transplantation.

The advances described here show how many of the challenges limiting RNA-guided therapies have been overcome. Some of the steps remain sub-optimal but advances are occurring at a rapid pace. Just as the Wright brothers' first demonstration of controlled flight paved the way for many advances that we now take for granted, the RNA-guided therapies promise a major change in the way we treat patients. The major hope is that we can use a standard chemistry to design guides that perform a highly specific function in cells, with easy checks to detect off-target effects, and outcomes optimized by redesign of the sequence or the use of different chemical modifications.

Throughout the book, examples are given of how we can use RNAs to down- and up-regulate gene expression, alter transcript processing through modulation of splicing, and recode proteins and modify their interactions with each other. These approaches do not alter the genome but instead reprogram the cell. They are possible because the logic of the cell is soft-wired. This logic is intransitive and implemented using direct cycles that change cell state as the context varies. Using RNA guides to reset those states and decrease the burden of disease offers much hope for the future. RNA therapeutics will become a commodity and a way to lower the cost of health care delivery for all.

Epilogue
Why Have a Career in Science?

The outcome of any set of experiments is always uncertain, despite all the preparation and careful analysis that precedes them. Even if the results confirm the initial hypothesis, they could still yield a false answer because of the bias subconsciously built into the experimental design. Even if you prove the result in many different ways, and the competing hypotheses are dismissed one after another, you have not yet arrived. The next hurdle is to publish your results and become known for your stellar work. If you are lucky, you will get to present your validated findings at a meeting and receive positive reviews. Usually, that is arranged by the club that has you as a member.

The hardest thing about science is the sense that nothing is ever quite complete. There is a control you didn't do, an interpretation you missed, a paper not published, manuscript not cited, a talk that went badly, an opportunity missed, time wasted on unfunded grant applications, and the difficult personalities along the way. At each step, you will have to answer for your decisions.

The best times are with the people you share the journey with. For me, the collaborations that stand out are where we left no stone unturned. It takes a while to develop the trust necessary to push on regardless of how the work unfolds. At the start, it is almost worth asking the simple question, "If I don't trust this person, why am I planning to work with them?". Inevitably, misunderstandings will arise that, hopefully, you will help to resolve quickly, as you already trust the other people involved.

At other times, I find it is best to be left alone – I would rather make a thousand mistakes in private rather than be constantly reminded of the one I made in public. For me, it's difficult to talk about a problem that I can't find the right words to describe. Usually, all the ideas emerge as a jumble. Unfortunately, the picture I vaguely see does not come pre-captioned. It usually takes me a while to find the best words as most of the image is initially quite blurry. If I am lucky, someone will be there to help me verbalize the images fluently. Initially, it is like a game of charades where they will try and guess the correct answer. Think of trying to explain an airplane to someone who has never seen one, but is rather skeptical of the idea. The conversation night unfold like this "You mean that it has wings like a bird? ... No, the wings don't flap ... So it's like a flying squirrel? ... No, it can go up and down and cover long distances. So, it floats in the air like a feather?... No, it's heavier than air..." and so on. During this process, it amuses me when I tell people that English is my second language and they believe me. Unfortunately, English is my only language. Even worse, when I explain my English difficulties by saying that I was dropped on my head as a child or it is a side effect of playing Rugby during my youth, they believe that as well! Now that I am older, I suspect that my English difficulties are attributed to incipient age-related dementia and no further explanation is necessary. Eventually, I find the few succinct words that will suffice.

For this reason, I do find the review process for manuscripts quite helpful. The responses help eliminate any ambiguities in the exposition. Here, the editor plays a critical role. Firstly, the editor needs to agree with you that the article is addressing a question of interest and that the manuscript is worthy of being sent out for review. Secondly, there is a chance that one reviewer will trash your work, so it is good to have an editor who can keep those comments in perspective. For me, I find that the vast majority of reviewers help with their comments and suggestions. Usually, they ask for additional figures or some explanations that you removed from the paper because of the word limitations specified by the journal guidelines. The reviewers will also ask for the additional references you left out because some journals limit those as well. The reviewers then obligate you to tie up those few loose ends and usually the editor will oblige and let you expand the paper. Lastly, a good editor may reject your paper, but still provide suggestions of another journal where there would be a better fit. They may also point out glaring pitfalls that you might want to address before the next submission. For all these reasons, there is often quite difference between my initial submission and the final publication. Much of the change has to do with my propensity to skip steps in the progression that are clear to me but are often obscure to others. Also, I am more motivated to put in the extra effort to perfect a paper once I know the paper will be published. Some of the initial submissions bounce around a bit until I find the right timbre. I still have a few preprints out there that I have been unable to progress.

Of course, there are editors who make their judgment of your paper based on impact factors that let them know how "hot" a field is at the moment. For Z-DNA, that meant a manuscript would not go out for review, even if really well written (see Figure 3.2). In fact, the Z-DNA scene was so bad at *Science* in 2021 that their editors did not seem to know that Z-DNA is a left-handed nucleic acid helix [211]. They published a paper about a right-handed helix containing a modified 2-diaminopurine base [212]. Of course, the reader by now knows that Z-DNA is left-handed. The authors, with the approval of the editors, referred to their right-handed structure as Z-DNA. They were playing on the abbreviation of 2-diaminopurine as "Z", just as adenosine is abbreviated as "A", so calling B-DNA containing the "Z" base Z-DNA made sense to them The *Science* staff should have caught the error. Initially, the editors did not think that the error was of any importance at all, refusing to publish a letter describing the issue as "unlikely to interest people outside the field of the paper" (letter to AH from Jennifer Stills, June 2, 2021). However, Rosalind Cotter from *Nature* indicated that they would publish a letter noting the mistake, commenting that "You make a valid point of clarification that is not contentious" (email to AH, June 3, 2021). So informed, *Science* persuaded the authors to rename their structure dZ-DNA, but without acknowledging their faux pas in the form of a published note or an erratum. In contrast, *Nature* did not hesitate to publish our letter pointing out the error (15 June 2021), while *Science* still declined to do so in print form. What can you do? To be fair, one editor of *Nature Genetics* recently said to me, when I pointed out that they had refused to send out for review my paper validating a genetic role for Z-DNA, "We do not do flipons". There is hope. At least that particular editor knows that flipons exist without me having to explain the concept to him. I'm not so sure about the current crop of editors at *Science*. An unfortunate aspect of these journals where the editors are not practicing scientists is their reliance on experts for an in-house review rather than sending the paper out to an external reviewer. There is no opportunity for an author to respond to the opinions voiced by the local expert as they are not shared. If the editor is forthright, as this editor from *Genome Research* was, the decision letter rejecting the manuscript might look like this":

"We therefore sought some informal advice from an expert in the field before making a decision; please note this is not a formal review, but a method by which we determine suitability of manuscripts where we feel some additional expertise is needed for the assessment" (email to author).

There you have it – a review by an expert without the right of an author's reply. Shades of the "double, secret, probation" from the movie *Animal House*, although without the mayhem. It happens.

It is great now that journals are offering the reviewers the choice of being named and acknowledged. I sign all my reviews. I think the anonymous reviewer is able to hide possible conflicts that may not be known to the editor. I am also amused by those nameless reviewers who draw attention to papers that should be cited. It seems, just possibly, that those papers could be ones the reviewer authored. It is relatively easy to make the connection when you actually suggested the person as a reviewer and they are actually the author of the paper that the reviewer says should be cited. Coincidence? You make the call. In the case of the first flipon paper, the reviewer, who mentioned her uncited paper, objected to many things, including the name "flipon". She stated that the field had not agreed upon the name. Which field was that I wondered? The one that had written off Z-DNA biology as a dead end? The reviewer actually helped with the publication of the paper as her review made it clear to the editor that the ideas expressed in the article were not settled science.

Also, I find it interesting that some scientists recommend each other to review their papers. This occurrence has been well documented. Clearly, there is the potential for a *quid pro quo* that undermines the peer review process. The argument against being named is that someone may pay you back for a negative review. However, if everyone is named, over time this problem takes care of itself. The particular biases of a reviewer become known to editors. If people refuse to participate in peer review, then that also will become known. Personally, I find reviewing takes a lot of time, especially involving papers with many authors and supplements as long, if not longer, than the article itself. I feel it is good to record my otherwise uncompensated contribution to science and also to pay back all the time that reviewers have put into my papers. If I am wrong, the author has a right to rebut the critiques, hopefully with better data. It is definitely a system that improves when openness is a core value.

Then, there are grant applications. Everything revolves about finding funding, which is based on the bedrock principle of peer review. Those peers are the ones who know all that there is to know about the known world you seek funding for. They therefore know the right things to fund, or at least they believe they know. So, if you are a know-it-all peer reviewer, why wouldn't you want to be known? Why is peer review anonymous? I have no good answer. All I know is that it is hard to push the limits of what is known if people don't seem to know what they don't know. If they don't know it can be done, how would they know if someone else might know how to do something they know only as impossible. Why just fund work that you know can be done and you know that the results will be what you know to expect? How do these anonymous know-it-alls know that this is the best that can be done? Apparently, because they know that, by being anonymous, no one will know whether or not they know what it is that they don't know.

Consequently, you will have your answer based on the decisions of others. Most of what happens then is out of your control. You may or may not get funding for the subsequent set of experiments or to finance collaborations with other scientists to develop your work further, or be invited to the speaker circuit or even receive an award. Such outcomes are uncertain and depend on a lot of non-scientific factors. The grant writing and the politics involved tend to remove you from the bench where your expertise once lay. Although you talk science, you don't actually do it anymore. The reality is that the experimental results you are selling are not your work, so make certain that those who are actually are doing the work receive the recognition they are due (see Figure 4.11).

Of course, getting noticed is part of a larger challenge. Will your paper be read or even cited correctly? According to the 2018 report of "The International Association of Scientific, Technical and Medical Publishers", there are 33,100 active scholarly peer-reviewed English-language journals collectively publishing over 3 million articles a year from 7–8 million researchers. It is a challenge to be noticed. The default for most people is to recite the standard papers in a field over and over again without necessarily having ever read them. The problem is further compounded by the many reviews of reviews where citations are amplified, even when they contain discredited claims or are no longer up to date. In fact, I am amused with the large number of reviews written nowadays. Many add to the author's *curriculum vitae* but nothing new. While this gives the reassuring appearance that the weight of evidence supports the opinions advanced, the contrafactual is that wrong ideas once established also reverberate through the literature just as well as those firmly supported by the evidence. Unfortunately, this makes it difficult to obtain funding to overturn that set of false beliefs, whose truth by the "field" is then held to be self-evident. Of course, I may be the one who is wrong. Just show me the data! If you don't have it, at least read the papers you are citing before referencing them in a review. Examples of false statements in reviews abound. In one recent paper, it was noted that Zα was found through a screen of a "chicken cDNA library for proteins that could bind to Z-DNA" [213]. I did ask the senior author of this paper, who is well known in the field, to correct the mistake by issuing an erratum, but so far, no response. I hope that such fading recollections will sunset into history and will not appear in subsequent reviews. If you spot such errors for yourself in any paper, it is a good hint that you should avoid working with those authors as it may be hard to catch their other, less obvious mistakes.

Your best response to these problems is through your own publications. Eventually, people will identify you with a body of work – what those accomplishments are is up to you. Then, you can address problems in the literature with your own findings, increasing the likelihood that your new data will become better known. Unfortunately, this strategy can require you to hyper-specialize in one area and that is all you are funded to do. In this age, NIH and NSF follow the Noah's Ark model and fund one, or preferably two, groups that have a specific expertise. If you are a junior investigator, that usually is not you. You might want to choose a field where one or both experts are due to retire. So, join the club and hang in there if you feel this option is best for funding your career.

Knowing what you don't know is important. It is necessary to optimize outcomes by questioning the limits of your knowledge. We constantly search for explanations as to why some things work and others don't. We then use the knowledge gained to predict future outcomes. As David Deutsch argues, good explanations lead to good predictions. Science quantitates results to see how well your ideas describe the future, putting aside the past as we have a tendency to select the evidence that supports what we know. We have already talked about the confirmation bias that paves the way to club membership. If you plan well, you will run experiments to optimize for good outcomes in areas where knowledge is poor, while avoiding the awkwardness of a posthumous Darwin Award for doing the things you should have known not to do. You can celebrate the former achievement, but not the latter. Science done properly does change the way we explain the world and our notion of what is possible.

We are always limited in what we can visualize both in our minds and with our instruments. Our views of life change with each technological advance. We knew little about cells until they became visible with the light microscope. Now, we can have tools to see large cellular machines in terms of the atoms that make them up. We can now see proteins that rotate like a wheel around an axis, being driven by positively charged hydrogen ions as they move from one side of a membrane to the other down a concentration gradient. In the process, the wheel's mechanical energy undergoes conversion into chemical energy. It drives the formation of adenosine triphosphate (ATP), paying for many of the molecular transformations that occur within a cell, with a phosphate offered up as compensation, a process first imagined by Peter Mitchell. With techniques capable of imaging single molecules, we can see at a resolution beyond half the wavelength of the light, once a feat considered so impossible that even Superman relied on X-ray vision. Newer approaches allow observation of single molecules in living cells by extremely powerful techniques that extract the signal from within the noise, The methods now enable tracking of molecules as they journey through a cell. Eventually, we will see flipons change conformation in real time and understand more about the causes and consequences of the flip.

The instruments available at a particular point in history limit the science that can be performed in each era. Once proved in principle by more primitive methods, and then put to practice using better apparatus, it is incredible how quickly the advances are made. DNA sequencing is but one example. Starting with a simple electrophoresis technique that allowed the 3.4 Å (10^{-10} of a meter) ladder of base pairs to be viewed with the naked eye 100 bases at a time (see Figure 4.10), we can now sequence billions of bases for a few hundred dollars, in a day or so. We are still not done with making these techniques even better as we can now sequence DNA one strand at a time through a nanopore.

While we have made great strides with cell biology, the big unknown is how the nervous system processes information to produce the conscious state. At present, there are no good explanations and the measurements we can now make on active brains in living subjects are of low resolution. But, just as the effects of lithium on mood and the effectiveness of anti-psychotics were unanticipated, I think the answers will emerge from approaches different from those currently in use. It will come down to making the right measurements and producing the best explanation for the results. We know the nervous system Is genetically encoded, but, just like the immune system, there are not enough genes in the genome to explain the variability in outcomes. We also know that, unlike the immune system, where cells can massively proliferate, the unit of selection is at the level of a synapse, rather than being cell-based. Like the immune system, the logic will be intransitive, going far beyond the current computational designs. Without intransitivity, nothing in biology would exist. Without curiosity, nothing new will be discovered. But without the right methods, progress will remain slow. Curiosity, methods and experimentation - there you have it. Dogma, elitism, and exclusion - then science loses its ability to perform magic.

There are many ways to pursue a career in science, rather than go it alone. One option is to be part of an empire like those well-funded institutes where resources are concentrated. These organizations can produce the plenteous pages of statistical tables of the type that please NIH program managers and procure the support of politicians who track the inflow of federal dollars into their districts. The papers published from these Institutes are in high-profile journals, albeit with a long list of authors. There are advantages to marching in step with the army, especially when no lethal weapons are involved. However, I note that these large institutions do burn a lot of money and they are dependent on funding that increasingly comes with strings attached. It is not a good sign when, as part of a small company, you reach out to see whether there is a possibility of a collaboration and the only two items discussed are the sponsored research agreement and the need to clear the subject area with the Institute's patent attorneys before any discussion can take place. There are many internal conflicts that arise at these organizations that limit scientific exchanges. Also, as Eric Lander, a founder of one large Institute in Cambridge, Massachusetts revealed, the leadership style at these places does not always translate well into the wider world (see an article by Alex Thompson in *Politico*, February 7, 2022).

I think the hardest choice is to partake in the competition for achieving the next milestone that everyone knows is coming. These situations involve three to five labs racing to verify the current expectations of how the world works, and, in the process, beat each other to press, by one means or another. The challenge is truly like a sporting event, akin to winning a World Championship – where one team takes the trophy. Just like with real live sport, it is certain that few will recall the winning score a year or so later or the names of the other teams in the play-offs. In contrast, people tend to remember who blew a critical play – think Bill Buckner and the Red Sox loss in the World Series. Reaching for the stars may sometimes end up with a crash landing, especially when others cannot reproduce the key findings of your high-profile paper or a co-author cannot locate the missing originals of a key figure.

Or you can try for the long shot and go for something that is currently just beyond the horizon. If successful, your papers will one day become cited as they will define the field, hopefully within your lifetime. Going long has more upside than a summer spent playing golf and a winter watching televised sports. Just don't expect a hole in one!

I have experienced many of these different styles. My joke about them all begins with the simple question I often ask: "How about …?" It was one of the things I liked about being at MIT. You would be talking through a project or something fairly random and an idea would come up, and so you would ask "How about we do this?" Usually, the immediate response would be "Yes, yes, that's a really great idea", followed by a moment of silence, then another, then another and then "No, no, I have a better way to do it" and so the conversation would continue. Later, when I moved to Boston University, the "How about …?" question would usually be followed by "Why?". Later still, when I went to Merck and asked the same question, the answer was "It's not your job".

There were other differences between MIT and Boston University, I was on the faculty at the Boston University Medical School where everything was to plan, like the army, one step at a time, cut-then-sew, tape then bandage. It was usual for people to be Head of Department for decades. Despite these impediments, I was able to advance human genetic studies using the data from the Framingham Heart Study. We were able to examine the entire human genome and find DNA variants that affected common traits like obesity and hypertension. It was an interesting culture clash, similar to the one I watched Jim Watson struggle with on his return to New Zealand. Molecular biologists like to publish fast while epidemiologists like to slowly accumulate their measurements and savor the implications and the accolades over a number of years. The joke about epidemiologists is that they are like Pharaohs, who would rather be buried with their data than share it. That thought always made me smile whenever I entered the office of the then-head of the Framingham Heart Study (FHS). The posts and lintel surrounding his office door had been enlarged with plaster to appear like the stone pillars that surround the opening of an ancient temple. The chain of command was no different from that of the army. You had to march to the right beat. Apparently, I didn't. On the political side, I failed because the science we did with Mike Christman was ahead of what the FHS investigators and the NIH reviewers of the study. Neither set of individuals thought the study was possible or that we were the ones remotely qualified to perform the work. I am OK with the outcome.

When I moved to Merck, hoping to translate genetic findings into therapies, the contrast with MIT was even more dramatic. Everyone had their job. When I think of my experience, the poem "The Charge of the Light Brigade" by Alfred, Lord Tennyson, surfaces.

"Theirs not to make reply,
Theirs not to reason why,
Theirs but to do and die".

Once a course of action was approved, you went through the mountain, not around it or over it. Tradition! There was no tolerance for "What if...?". Understandably, I was there for only a short time, yet, despite this, a program I worked on went from basic science to the clinic and another one I helped to initiate continued well after I left. At both Boston University and Merck, just like the Army, I stepped out of line and was handed my "boots". The big difference was the amount of cash that the separation from Merck came with. Both were interesting experiences and I learned many new things that helped me later on. On the plus side, the pay and benefits at both of these institutions were much better than at MIT! On the minus side, as I walked cut the door of Merck, any good science I accomplished was left behind locked up in their proprietary vault with my former colleagues immersed in a silo of silence.

After exiting Merck, having failed the personal improvement program, separation check in hand, I came back to the Z-DNA question. I was disappointed that although the Z-DNA binding domain I discovered was still mentioned in the literature, it was always described as being "of unknown function". There was a job to finish. This book is about how the science wins out in the end, even when you take the long odds and even when the smart money is not on you. The surprising thing to me was that I was able to achieve so much without an academic affiliation. The accomplishments were only possible because of the great collaborators I worked with. Each made the other better. The only negative thing about being without an official-sounding title was that some news aggregator like Science X would not feature my publications since they only sourced their feeds from "a research institution, university, or a respectable scientific organization". (email to author). Apparently, I keep finding ways to be rejected as "none of the above".

More often than not, the best outcome of these endeavors is to survive to fight another day. There is still a lot to discover and a lot to comprehend. The aspect that has already amazed me most about biology is how few things are well understood. It doesn't take too long to find the limits of knowledge and important questions to ask and hopefully answer. Of course, as I found with my medical school professors, daring to ask a question expecting an answer based on well-controlled experiments can give the impression that you are not being respectful of their authority. That can get you into trouble. I also found that there is also a very fine line between being flippant and appearing sarcastic. Saying "You're joking, aren't you?" to someone you know does not have a sense of humor is definitely not a good move.

You will find in your career that, when various issues arise, when lines are crossed and you when you doubt your choices, it is often better to walk away. This is the same strategy used by the union guys I worked with during my student days. There is the principle of *quid pro quo* that underlies those decisions. Whether to fight or not is your choice. I find it more satisfying to focus that energy on the outcomes I have some control over. It is my observation that things generally work out. It does not pay to double down on losing bets. Those who make bad or mean-spirited decisions that adversely affect you tend to do the same thing to others. Their behavior does catch up with them. Play the long game rather than get involved in a tit-for-tat where the chances of you winning the battle are much less than the probability of you eventually finding success, regardless of how unlikely that looks at a particular moment in time. A good outcome depends on the help of others and more than a measure of luck. It depends on the path you choose. No one else will be walking in your boots. Unfortunately, you will have to deal with the sergeants of science who expect you to march a short step. If you prefer the pageantry of the parade ground, then someone will certainly be there to make sure that the commanding officer is impressed by the shine on your shoes.

Be mindful that our work is made possible by many others. We all do science to make a difference. The opportunities are only there because of what others did before. Hopefully, you will do the same for those who follow. We do depend on the support of our communities and the enthusiastic participation of families, such as those from Framingham. Our discoveries hopefully will repay their trust. Science has a proud tradition of closing the gap between the present and the future, while striving to benefit all. Some arrive before others leave, but that's just the way change happens.

So why have a career in science? It's the only way to find out whether something you imagine is truly impossible, and even then, you may not know for sure until you ask the right question. Even smart people are sometimes wrong. So, follow the data, use the methods most appropriate, and publish what you find. Be truthful with yourself and with others. Otherwise, the only person you will fool is yourself. You can learn much from how a person responds to being asked "How about we ...?" or "What if...?". Paying forward rather than paying back will keep your dream alive. Most of all, be lucky enough to find collaborators willing to chance it all, knowing that challenges, critics, and competition are certain, whereas success is not.

References

1. Stigler SM. *Statistics on the table: The history of statistical concepts and methods.* Cambridge, MA: Harvard University Press; 1999.
2. Crick F. Split genes and RNA splicing. *Science.* 1979 Apr 20;204(4390):264–71.
3. Herbert A. To "Z" or not to "Z": Z-RNA, self-recognition, and the MDA5 helicase. *PLoS Genet.* 2021 May;17(5):e1009513.
4. Watson JD, Crick FH. Molecular structure of nucleic acids; A structure for deoxyribose nucleic acid. *Nature.* 1953 Apr 25;171(4356):737–8.
5. Wilkins MH, Stokes AR, Wilson HR. Molecular structure of deoxypentose nucleic acids. *Nature.* 1953 Apr 25;171(4356):738–40.
6. Franklin RE, Gosling RG. The structure of sodium thymonucleate fibres. I. The influence of water content. *Acta Crystallographica.* 1953;6(8):673–7.
7. Perutz MF, Randall JT, Thomson L, et al. DNA helix. *Science.* 1969 Jun 27;164(3887):1537–9.
8. Hargittai I. The tetranucleotide hypothesis: A centennial. *Structural Chemistry.* 2009;20(5):753–6.
9. Kay LE. W. M. Stanley's crystallization of the tobacco mosaic virus, 1930–1940. *Isis.* 1986 Sep;77(288):450–72.
10. Avery OT, Macleod CM, McCarty M. Studies on the chemical nature of the substance inducing transformation of pneumococcal types: Induction of transformation by a desoxyribonucleic acid fraction isolated from pneumococcus type III. *J Exp Med.* 1944 Feb 1;79(2):137–58.
11. Olby RC. *The path to the double helix: The discovery of DNA.* New York: Dover Publications; 1994.
12. Nanjundiah V. George Gamow and the genetic code. *Resonance.* 2004;9(7):44–9.
13. Wiener N. *Cybernetics.* New York: J. Wiley; 1948.
14. Jacob F, Monod J. Genetic regulatory mechanisms in the synthesis of proteins. *J Mol Biol.* 1961 Jun;3:318–56.
15. Baltimore D. RNA-dependent DNA polymerase in virions of RNA tumour viruses. *Nature.* 1970 Jun 27;226(5252):1209–11.
16. Temin HM, Mizutani S. RNA-dependent DNA polymerase in virions of Rous sarcoma virus. *Nature.* 1970 Jun 27;226(5252):1211–3.
17. Chow LT, Gelinas RE, Broker TR, et al. An amazing sequence arrangement at the 5′ ends of adenovirus 2 messenger RNA. *Cell.* 1977 Sep;12(1):1–8.
18. Berget SM, Moore C, Sharp PA. Spliced segments at the 5′ terminus of adenovirus 2 late mRNA. *Proc Natl Acad Sci U S A.* 1977 Aug;74(8):3171–5.
19. Smith HH, Brookhaven National Laboratory., U.S. Atomic Energy Commission. *Evolution of genetic systems.* New York: Gordon and Breach; 1972.
20. Herbert A. The four Rs of RNA-directed evolution. *Nat Genet.* 2004 Jan;36(1):19–25.
21. Brenner S. Refuge of spandrels. *Curr Biol.* 1998 Sep 24;8(19):R669.
22. Macrae AD, Brenner S. Analysis of the dopamine receptor family in the compact genome of the puffer fish Fugu rubripes. *Genomics.* 1995 Jan 20;25(2):436–46.
23. Drew HR, Dickerson RE, Itakura K. A salt-induced conformational change in crystals of the synthetic DNA tetramer d(CpGpCpG). *J Mol Biol.* 1978 Nov 15;125(4):535–43.
24. Wang AH, Quigley GJ, Kolpak FJ, et al. Molecular structure of a left-handed double helical DNA fragment at atomic resolution [Research Support, U.S. Gov't, P.H.S.]. *Nature.* 1979 Dec 13;282(5740):680–6.

25. Pohl FM, Jovin TM. Salt-induced co-operative conformational change of a synthetic DNA: Equilibrium and kinetic studies with poly (dG-dC). *J Mol Biol.* 1972 Jun 28;67(3):375–96.

26. Santella RM, Grunberger D, Broyde S, et al. Z-DNA conformation of N-2-acetylaminofluorene modified poly(dG-dC).poly(dG-dC) determined by reactivity with anti cytidine antibodies and minimized potential energy calculations. *Nucleic Acids Res.* 1981 Oct 24;9(20):5459–67.

27. Moller A, Nordheim A, Kozlowski SA, et al. Bromination stabilizes poly(dG-dC) in the Z-DNA form under low-salt conditions. *Biochemistry.* 1984 Jan 3;23(1):54–62.

28. Lafer EM, Valle RP, Moller A, et al. Z-DNA-specific antibodies in human systemic lupus erythematosus. *J Clin Invest.* 1983 Feb;71(2):314–21.

29. Singleton CK, Klysik J, Stirdivant SM, et al. Left-handed Z-DNA is induced by supercoiling in physiological ionic conditions. *Nature.* 1982 Sep 23;299(5881):312–6.

30. Peck LJ, Wang JC. Energetics of B-to-Z transition in DNA [Research support, U.S. Gov't, Non-P.H.S.]. *Proc Natl Acad Sci U S A.* 1983 Oct;80(20):6206–10.

31. Morange M. What history tells us IX. Z-DNA: When nature is not opportunistic. *J Biosci.* 2007 Jun;32(4):657–61.

32. Hozumi N, Tonegawa S. Evidence for somatic rearrangement of immunoglobulin genes coding for variable and constant regions. *Proc Natl Acad Sci U S A.* 1976 Oct;73(10):3628–32.

33. Pauling L, Corey RB. A proposed structure for the nucleic acids. *Proc Natl Acad Sci U S A.* 1953 Feb;39(2):84–97.

34. Schimmel P. Alexander Rich (1924–2015). *Nature.* 2015 May 21;521(7552):291.

35. Kim SH, Suddath FL, Quigley GJ, et al. Three-dimensional tertiary structure of yeast phenylalanine transfer RNA. *Science.* 1974 Aug 2;185(4149):435–40.

36. Robertus JD, Ladner JE, Finch JT, et al. Structure of yeast phenylalanine tRNA at 3 A resolution. *Nature.* 1974 Aug 16;250(467):546–51.

37. Zhang S, Holmes T, Lockshin C, et al. Spontaneous assembly of a self-complementary oligopeptide to form a stable macroscopic membrane. *Proc Natl Acad Sci U S A.* 1993 Apr 15;90(8):3334–8.

38. Herbert AG, Rich A. A method to identify and characterize Z-DNA binding proteins using a linear oligodeoxynucleotide [Research support, Non-U.S. Gov't research support, U.S. Gov't, Non-P.H.S. Research support, U.S. Gov't, P.H.S.]. *Nucleic Acids Res.* 1993 Jun 11;21(11):2669–72.

39. Herbert AG, Spitzner JR, Lowenhaupt K, et al. Z-DNA binding protein from chicken blood nuclei [Research support, Non-U.S. Gov't research support, U.S. Gov't, Non-P.H.S. Research support, U.S. Gov't, P.H.S.]. *Proc Natl Acad Sci U S A.* 1993 Apr 15;90(8):3339–42.

40. Herbert A, Lowenhaupt K, Spitzner J, et al. Chicken double-stranded RNA adenosine deaminase has apparent specificity for Z-DNA [Comparative study]. *Proc Natl Acad Sci U S A.* 1995 Aug 1;92(16):7550–4.

41. Kim U, Wang Y, Sanford T, et al. Molecular cloning of cDNA for double-stranded RNA adenosine deaminase, a candidate enzyme for nuclear RNA editing. *Proc Natl Acad Sci U S A.* 1994 Nov 22;91(24):11457–61.

42. Bass BL, Weintraub H. An unwinding activity that covalently modifies its double-stranded RNA substrate. *Cell.* 1988 Dec 23;55(6):1089–98.

43. Herbert A, Alfken J, Kim YG, et al. A Z-DNA binding domain present in the human editing enzyme, double-stranded RNA adenosine deaminase [Research support, Non-U.S. Gov't research support, U.S. Gov't, Non-P.H.S. Research support, U.S. Gov't, P.H.S.]. *Proc Natl Acad Sci U S A.* 1997 Aug 5;94(16):8421–6.

44. Schade M, Turner CJ, Lowenhaupt K, et al. Structure-function analysis of the Z-DNA-binding domain Zα of dsRNA adenosine deaminase type I reveals similarity to the (alpha + beta) family of helix-turn-helix proteins [Research support, Non-U.S. Gov't research support, U.S. Gov't, Non-P.H.S. Research support, U.S. Gov't, P.H.S.] *Embo J.* 1999 Jan 15;18(2):470–9.

45. Herbert A, Schade M, Lowenhaupt K, et al. The Zα domain from human ADAR1 binds to the Z-DNA conformer of many different sequences. *Nucleic Acids Res.* 1998 Aug 1;26(15):3486–93.

46. Kim YG, Kim PS, Herbert A, et al. Construction of a Z-DNA-specific restriction endonuclease. *Proc Natl Acad Sci U S A.* 1997 Nov 25;94(24):12875–9.

47. Schwartz T, Rould MA, Lowenhaupt K, et al. Crystal structure of the Zα domain of the human editing enzyme ADAR1 bound to left-handed Z-DNA [Research support, U.S. Gov't, Non-P.H.S. Research support, U.S. Gov't, P.H.S.]. *Science.* 1999 Jun 11;284(5421):1841–5.

48. Schade M, Turner CJ, Kuhne R, et al. The solution structure of the Zα domain of the human RNA editing enzyme ADAR1 reveals a prepositioned binding surface for Z-DNA [Research support, Non-U.S. Gov't research support, U.S. Gov't, Non-P.H.S. Research support, U.S. Gov't, P.H.S.]. *Proc Natl Acad Sci U S A.* 1999 Oct 26;96(22):12465–70.

49. Higuchi M, Maas S, Single FN, et al. Point mutation in an AMPA receptor gene rescues lethality in mice deficient in the RNA-editing enzyme ADAR2. *Nature.* 2000 Jul 6;406(6791):78–81.

50. Herbert A, Rich A. The role of binding domains for dsRNA and Z-DNA in the in vivo editing of minimal substrates by ADAR1 [Research support, U.S. Gov't, Non-P.H.S. Research support, U.S. Gov't, P.H.S.]. *Proc Natl Acad Sci U S A.* 2001 Oct 9;98(21):12132–7.

51. Van Steen K, McQueen MB, Herbert A, et al. Genomic screening and replication using the same data set in family-based association testing. *Nat Genet.* 2005 Jul;37(7): 683–91.

52. Herbert A, Gerry NP, McQueen MB, et al. A common genetic variant is associated with adult and childhood obesity. *Science.* 2006 Apr 14;312(5771):279–83.

53. Kang SP, Gergich K, Lubiniecki GM, et al. Pembrolizumab KEYNOTE-001: An adaptive study leading to accelerated approval for two indications and a companion diagnostic. *Ann Oncol.* 2017 Jun 1;28(6):1388–98.

54. Cho MS, Vasquez HG, Rupaimoole R, et al. Autocrine effects of tumor-derived complement. *Cell Rep.* 2014 Mar 27;6(6):1085–95.

55. Herbert A. Complement controls the immune synapse and tumors control complement. *J Immunother Cancer.* 2020;8(2):e001712. doi: 10.1136/jitc-2020-001712. PubMed PMID: 33323465; PubMed Central PMCID: PMC7745530.

56. Editorial. The case for lotteries as a tiebreaker of quality in research funding. *Nature.* 2022 Sep;609(7928):653.

57. Ha SC, Lowenhaupt K, Rich A, et al. Crystal structure of a junction between B-DNA and Z-DNA reveals two extruded bases [Research support, N.I.H., Extramural research support, Non-U.S. Gov't research support, U.S. Gov't, P.H.S.]. *Nature.* 2005 Oct 20;437(7062):1183–6.

58. Placido D, Brown BA, Lowenhaupt K, et al. A left-handed RNA double helix bound by the Z alpha domain of the RNA-editing enzyme ADAR1 [Research support, N.I.H., Extramural research support, Non-U.S. Gov't]. *Structure.* 2007 Apr;15(4): 395–404.

59. Kim YG, Muralinath M, Brandt T, et al. A role for Z-DNA binding in vaccinia virus pathogenesis. *Proc Natl Acad Sci U S A.* 2003 Jun 10;100(12):6974–9.

60. Kwon JA, Rich A. Biological function of the vaccinia virus Z-DNA-binding protein E3L: Gene transactivation and antiapoptotic activity in HeLa cells. *Proc Natl Acad Sci U S A*. 2005 Sep 6;102(36):12759–64.

61. Athanasiadis A. Zα-domains: At the intersection between RNA editing and innate immunity. *Seminars in Cell & Developmental Biology*. 2012;23(3):275–80.

62. Wang Q, Khillan J, Gadue P, et al. Requirement of the RNA editing deaminase ADAR1 gene for embryonic erythropoiesis [Research support, Non-U.S. Gov't research support, U.S. Gov't, P.H.S.]. *Science*. 2000 Dec 1;290(5497):1765–8.

63. Hartner JC, Schmittwolf C, Kispert A, et al. Liver disintegration in the mouse embryo caused by deficiency in the RNA-editing enzyme ADAR1. *J Biol Chem*. 2004 Feb 6;279(6):4894–902.

64. Hartner JC, Walkley CR, Lu J, et al. ADAR1 is essential for the maintenance of hematopoiesis and suppression of interferon signaling. *Nat Immunol*. 2009 Jan;10(1):109–15.

65. Ward SV, George CX, Welch MJ, et al. RNA editing enzyme adenosine deaminase is a restriction factor for controlling measles virus replication that also is required for embryogenesis [Research support, N.I.H., Extramural]. *Proc Natl Acad Sci U S A*. 2011 Jan 4;108(1):331–6.

66. Liddicoat BJ, Piskol R, Chalk AM, et al. RNA editing by ADAR1 prevents MDA5 sensing of endogenous dsRNA as nonself [Research Support, N.I.H., Extramural research support, Non-U.S. Gov't]. *Science*. 2015 Sep 4;349(6252):1115–20.

67. Levanon EY, Eisenberg E, Yelin R, et al. Systematic identification of abundant A-to-I editing sites in the human transcriptome [Comparative study evaluation studies research support, Non-U.S. Gov't validation studies]. *Nat Biotechnol*. 2004 Aug;22(8):1001–5.

68. Kim DD, Kim TT, Walsh T, et al. Widespread RNA editing of embedded alu elements in the human transcriptome [Research support, U.S. Gov't, P.H.S.]. *Genome Res*. 2004 Sep;14(9):1719–25.

69. Blow M, Futreal PA, Wooster R, et al. A survey of RNA editing in human brain [Comparative study]. *Genome Res*. 2004 Dec;14(12):2379–87.

70. Athanasiadis A, Rich A, Maas S. Widespread A-to-I RNA editing of Alu-containing mRNAs in the human transcriptome [Research support, Non-U.S. Gov't]. *PLoS Biol*. 2004 Dec;2(12):e391.

71. Herbert A. ALU non-B-DNA conformations, flipons, binary codes and evolution. *Royal Society Open Science*. 2020;7(6):200222.

72. Herbert A. Z-DNA and Z-RNA in human disease. *Communications Biology*. 2019 Jul 1;2(1):7.

73. Barak M, Porath HT, Finkelstein G, et al. Purifying selection of long dsRNA is the first line of defense against false activation of innate immunity. *Genome Biol*. 2020 Feb 7;21(1):26.

74. Nichols PJ, Bevers S, Henen M, et al. Recognition of non-CpG repeats in Alu and ribosomal RNAs by the Z-RNA binding domain of ADAR1 induces A-Z junctions. *Nat Commun*. 2021;12(1):793.

75. Rice GI, Kasher PR, Forte GM, et al. Mutations in ADAR1 cause Aicardi-Goutieres syndrome associated with a type I interferon signature [Research support, N.I.H., Extramural research support, Non-U.S. Gov't]. *Nat Genet*. 2012 Nov;44(11):1243–8.

76. Livingston JH, Lin JP, Dale RC, et al. A type I interferon signature identifies bilateral striatal necrosis due to mutations in ADAR1 [Research support, Non-U.S. Gov't]. *J Med Genet*. 2014 Feb;51(2):76–82.

77. Herbert A. Mendelian disease caused by variants affecting recognition of Z-DNA and Z-RNA by the Zα domain of the double-stranded RNA editing enzyme ADAR. *Eur J Hum Genet*. 2020 Jan;28(1):114–17.

78. Hayashi M, Suzuki T. Dyschromatosis symmetrica hereditaria [Review]. *J Dermatol.* 2013 May;40(5):336–43.

79. Guo X, Liu S, Sheng Y, et al. ADAR1 Zalpha domain P195A mutation activates the MDA5-dependent RNA-sensing signaling pathway in brain without decreasing overall RNA editing. *Cell Rep.* 2023 Jul 25;42(7):112733.

80. Liang Z, Chalk AM, Taylor S, et al. The phenotype of the most common human ADAR1p150 Zalpha mutation P193A in mice is partially penetrant. *EMBO Rep.* 2023 May 4;24(5):e55835.

81. Herbert A. ADAR and immune silencing in cancer. *Trends Cancer.* 2019 May;5(5):272–82.

82. Ishizuka JJ, Manguso RT, Cheruiyot CK, et al. Loss of ADAR1 in tumours overcomes resistance to immune checkpoint blockade. *Nature.* 2019 Jan;565(7737):43–48

83. Behan FM, Iorio F, Picco G, Goncalves E, Beaver CM, Migliardi G, et al. Prioritization of cancer therapeutic targets using CRISPR-Cas9 screens. *Nature.* 2019;568(7753):511–6. Epub 20190410. doi: 10.1038/s41586-019-1103-9. PubMed PMID: 30971826.

84. Fu Y, Comella N, Tognazzi K, et al. Cloning of DLM-1, a novel gene that is up-regulated in activated macrophages, using RNA differential display. *Gene.* 1999 Nov 15;240(1):157–63.

85. Schwartz T, Behlke J, Lowenhaupt K, et al. Structure of the DLM-1-Z-DNA complex reveals a conserved family of Z-DNA-binding proteins. *Nat Struct Biol.* 2001 Sep;8(9):761–5.

86. Takaoka A, Wang Z, Choi MK, et al. DAI (DLM-1/ZBP1) is a cytosolic DNA sensor and an activator of innate immune response. *Nature.* 2007;448(7152):501–5.

87. Upton JW, Kaiser WJ, Mocarski ES. DAI/ZBP1/DLM-1 complexes with RIP3 to mediate virus-induced programmed necrosis that is targeted by murine cytomegalovirus vIRA. *Cell Host Microbe.* 2012 Mar 15;11(3):290–7.

88. Kuriakose T, Man SM, Malireddi RK, Karki R, Kesavardhana S, Place DE, et al. ZBP1/DAI is an innate sensor of influenza virus triggering the NLRP3 inflammasome and programmed cell death pathways. *Sci Immunol.* 2016 Aug 5;1(2):aag2045. Epub 20160812. doi: 10.1126/sciimmunol.aag2045. PubMed PMID: 27917412; PubMed Central PMCID: PMC5131924.

89. Rebsamen M, Heinz LX, Meylan E, et al. DAI/ZBP1 recruits RIP1 and RIP3 through RIP homotypic interaction motifs to activate NF-kappaB. *EMBO Rep.* 2009 Aug;10(8):916–22.

90. Kaiser WJ, Upton JW, Mocarski ES. Receptor-interacting protein homotypic interaction motif-dependent control of NF-kappa B activation via the DNA-dependent activator of IFN regulatory factors. *J Immunol.* 2008 Nov 1;181(9):6427–34.

91. Thapa RJ, Ingram JP, Ragan KB, et al. DAI senses influenza a virus genomic RNA and activates RIPK3-dependent cell death. *Cell Host Microbe.* 2016;20(5):674–81.

92. Zhang T, Yin C, Boyd DF, et al. Influenza virus Z-RNAs Induce ZBP1-mediated necroptosis. *Cell.* 2020 Mar 19;180(6):1115–29.

93. Sridharan H, Ragan KB, Guo H, et al. Murine cytomegalovirus IE3-dependent transcription is required for DAI/ZBP1-mediated necroptosis [Research support, Non-U.S. Gov't research support, N.I.H., Extramural]. *EMBO Rep.* 2017 Aug;18(8):1429–41.

94. Maelfait J, Liverpool L, Bridgeman A, et al. Sensing of viral and endogenous RNA by ZBP1/DAI induces necroptosis [Research support, Non-U.S. Gov't]. *Embo J.* 2017 Sep 1;36(17):2529–43.

95. Nassour J, Aguiar LG, Correia A, et al. Telomere-to-mitochondria signalling by ZBP1 mediates replicative crisis. *Nature.* 2023 Feb;614(7949):767–73.

96. Lei Y, VanPortfliet JJ, Chen YF, Bryant JD, Li Y, Fails D, et al. Cooperative sensing of mitochondrial DNA by ZBP1 and cGAS promotes cardiotoxicity. *Cell.* 2023;186(14):3013–32 e22. Epub 20230622. doi: 10.1016/j.cell.2023.05.039. PubMed PMID: 37352855; PubMed Central PMCID: PMC10330843.

97. Koehler H, Cotsmire S, Zhang T, et al. Vaccinia virus E3 prevents sensing of Z-RNA to block ZBP1-dependent necroptosis. *Cell Host Microbe.* 2021 Jun 24;29:1266–76.e5.

98. Sun L, Miao Y, Wang Z, Chen H, Dong P, Zhang H, et al. Structural insight into African swine fever virus I73R protein reveals it as a Z-DNA binding protein. *Transbound Emerg Dis.* 2022;69(5):e1923–e35. Epub 20220404. doi: 10.1111/tbed.14527. PubMed PMID: 35312168.

99. Marshall PR, Zhao Q, Li X, Wei W, Periyakaruppiah A, Zajaczkowski EL, et al. Dynamic regulation of Z-DNA in the mouse prefrontal cortex by the RNA-editing enzyme Adar1 is required for fear extinction. *Nat Neurosci.* 2020;23(6):718–29. Epub 20200504. doi: 10.1038/s41593-020-0627-5. PubMed PMID: 32367065; PubMed Central PMCID: PMC7269834.

100. Pardue ML, Nordheim A, Lafer EM, et al. Z-DNA and the polytene chromosome. *Cold Spring Harb Symp Quant Biol.* 1983;47 Pt 1:171–6.

101. Arndt-Jovin DJ, Robert-Nicoud M, Zarling DA, et al. Left-handed Z-DNA in bands of acid-fixed polytene chromosomes. *Proc Natl Acad Sci U S A.* 1983 Jul;80(14):4344–8.

102. Hill RJ, Stollar BD. Dependence of Z-DNA antibody binding to polytene chromosomes on acid fixation and DNA torsional strain. *Nature.* 1983;305(5932):338–40.

103. Zarling DA, Calhoun CJ, Hardin CC, et al. Cytoplasmic Z-RNA. *Proc Natl Acad Sci U S A.* 1987 Sep;84(17):6117–21.

104. Zhang T, Yin C, Fedorov A, et al. ADAR1 masks the cancer immunotherapeutic promise of ZBP1-driven necroptosis. *Nature.* 2022 Jun;606(7914):594–602.

105. George CX, Ramaswami G, Li JB, et al. Editing of cellular self-RNAs by adenosine deaminase ADAR1 suppresses innate immune stress responses [research support, N.I.H., Extramural research support, Non-U.S. Gov't]. *J Biol Chem.* 2016 Mar 18;291(12):6158–68.

106. Liu Y, Lei M, Samuel CE. Chimeric double-stranded RNA-specific adenosine deaminase ADAR1 proteins reveal functional selectivity of double-stranded RNA-binding domains from ADAR1 and protein kinase PKR. *Proc Natl Acad Sci U S A.* 2000 Nov 7;97(23):12541–6.

107. Sarantopoulos J, Mahalingam D, Sharma N, et al. Results of a completed phase I trial of CBL0137 administered intravenously (IV) to patients (Pts) with advanced solid tumors. *Journal of Clinical Oncology.* 2020;38(15_suppl):3583.

108. Herbert A, Balachandran S. Z-DNA enhances immunotherapy by triggering death of inflammatory cancer-associated fibroblasts. *J Immunother Cancer.* 2022;10(11):e005704. doi: 10.1136/jitc-2022-005704. PubMed PMID: 36450382; PubMed Central PMCID: PMC9716847.

109. Jiao H, Wachsmuth L, Wolf S, et al. ADAR1 averts fatal type I interferon induction by ZBP1. *Nature.* 2022 Jul;607(7920):776–83.

110. Hubbard NW, Ames JM, Maurano M, et al. ADAR1 mutation causes ZBP1-dependent immunopathology. *Nature.* 2022 Jul;607(7920):769–75.

111. de Reuver R, Verdonck S, Dierick E, et al. ADAR1 prevents autoinflammation by suppressing spontaneous ZBP1 activation. *Nature.* 2022 Jul;607(7920):784–9.

112. Karki R, Sundaram B, Sharma BR, et al. ADAR1 restricts ZBP1-mediated immune response and PANoptosis to promote tumorigenesis. *Cell Rep.* 2021 Oct 19;37(3):109858.

113. Buzzo JR, Devaraj A, Gloag ES, et al. Z-form extracellular DNA is a structural component of the bacterial biofilm matrix. *Cell.* 2021 Nov 11;184(23):5740–58 e17.

114. Seviour T, Winnerdy FR, Wong LL, et al. The biofilm matrix scaffold of Pseudomonas aeruginosa contains G-quadruplex extracellular DNA structures. *NPJ Biofilms Microbiomes.* 2021 Mar 19;7(1):27.

115. Iwaniuk EE, Adebayo T, Coleman S, et al. Activatable G-quadruplex based catalases for signal transduction in biosensing. *Nucleic Acids Res.* 2023 Feb 28;51(4):1600–7.

116. Herbert A, Fedorov A, Poptsova M. Mono a Mano: ZBP1's Love-Hate Relationship with the Kissing Virus. *Int J Mol Sci.* 2022 Mar 12;23(6):3079.

117. Harley JB, Chen X, Pujato M, et al. Transcription factors operate across disease loci, with EBNA2 implicated in autoimmunity. *Nat Genet.* 2018 May;50(5):699–707

118. Minero GAS, Møllebjerg A, Thiesen C, Johansen MI, Jorgensen NP, Birkedal V, et al. Extracellular G-quadruplexes and Z-DNA protect biofilms from DNase I, and G-quadruplexes form a DNAzyme with peroxidase activity. *Nucleic Acids Res.* 2024. Epub 20240131. doi: 10.1093/nar/gkae034. PubMed PMID: 38296834.

119. Ho PS, Ellison MJ, Quigley GJ, et al. A computer aided thermodynamic approach for predicting the formation of Z-DNA in naturally occurring sequences [Research Support, Non-U.S. Gov't Research Support, U.S. Gov't, Non-P.H.S. Research Support, U.S. Gov't, P.H.S.]. *Embo J.* 1986 Oct;5(10):2737–44.

120. Benham CJ. Theoretical analysis of transitions between B- and Z-conformations in torsionally stressed DNA. *Nature.* 1980 Aug 7;286(5773):637–8.

121. Wittig B, Dorbic T, Rich A. The level of Z-DNA in metabolically active, permeabilized mammalian cell nuclei is regulated by torsional strain. *J Cell Biol.* 1989 Mar;108(3):755–64.

122. Wittig B, Wolfl S, Dorbic T, et al. Transcription of human c-myc in permeabilized nuclei is associated with formation of Z-DNA in three discrete regions of the gene [Research support, Non-U.S. Gov't research support, U.S. Gov't, Non-P.H.S. Research support, U.S. Gov't, P.H.S.]. *Embo J.* 1992 Dec;11(12):4653–63.

123. Kouzine F, Wojtowicz D, Baranello L, et al. Permanganate/S1 nuclease footprinting reveals Non-B DNA structures with regulatory potential across a mammalian genome. *Cell Syst.* 2017 Mar 22;4(3):344–56.

124. Umerenkov D, Herbert A, Konovalov D, Danilova A, Beknazarov N, Kokh V, et al. Z-flipon variants reveal the many roles of Z-DNA and Z-RNA in health and disease. *Life Sci Alliance.* 2023;6(7):e202301962. Epub 20230510. doi: 10.26508/lsa.202301962. PubMed PMID: 37164635; PubMed Central PMCID: PMC10172764.

125. Herbert A. The Intransitive Logic of Directed Cycles and Flipons Enhances the Evolution of Molecular Computers by Augmenting the Kolmogorov Complexity of Genomes. *International Journal of Molecular Sciences.* 2023;24(22):16482. doi: 10.3390/ijms242216482.

126. Herbert A. Nucleosomes and flipons exchange energy to alter chromatin conformation, the readout of genomic information, and cell fate. *Bioessays.* 2022;44(12):e2200166. Epub 20221101. doi: 10.1002/bies.202200166. PubMed PMID: 36317523.

127. Liu H, Mulholland N, Fu H, et al. Cooperative activity of BRG1 and Z-DNA formation in chromatin remodeling. *Mol Cell Biol.* 2006 Apr;26(7):2550–9.

128. Maruyama A, Mimura J, Harada N, et al. Nrf2 activation is associated with Z-DNA formation in the human HO-1 promoter [Research support, Non-U.S. Gov't]. *Nucleic Acids Res.* 2013 May 1;41(10):5223–34.

129. Bartas M, Slychko K, Cerven J, Pecinka P, Arndt-Jovin DJ, Jovin TM. Extensive Bioinformatics Analyses Reveal a Phylogenetically Conserved Winged Helix (WH) Domain (Ztau) of Topoisomerase IIalpha, Elucidating Its Very High Affinity for Left-Handed Z-DNA and Suggesting Novel Putative Functions. *Int J Mol Sci.* 2023;24(13):10740. Epub 20230627. doi: 10.3390/ijms241310740. PubMed PMID: 37445918; PubMed Central PMCID: PMC10341724.

130. Hui J, Hung L-H, Heiner M, et al. Intronic CA-repeat and CA-rich elements: A new class of regulators of mammalian alternative splicing. *Embo J.* 2005;24(11):1988–98.

131. Jiang D, Zhang J. The preponderance of nonsynonymous A-to-I RNA editing in coleoids is nonadaptive. *Nat Commun.* 2019 Nov 27;10(1):5411.

132. Packalen M, Bhattacharya J. NIH funding and the pursuit of edge science. *Proc Natl Acad Sci U S A.* 2020 Jun 2;117(22):12011–16.

133. Stent GS. Prematurity and uniqueness in scientific discovery. *Scientific American.* 1972;227(6):84–93.

134. Player G. *Gary player's golf secrets.* Englewood Cliffs, NJ: Prentice-Hall; 1962.

135. Jackson JD. Examples of the zeroth theorem of the history of science. *American Journal of Physics.* 2008;76(8):704–19.

136. Saltz A. The true story of the discovery of streptomycin. *Actinomycetes.* 1993;4(2): 27–39.

137. Darwin F. *The eugenics review.* Eugenics Education Society; 1914, Vol. 6.

138. Abir-Am P. The women who discovered RNA splicing. *American Scientist.* 2020;108:298.

139. Glover DM, Hogness DS. A novel arrangement of the 18S and 28S sequences in a repeating unit of Drosophila melanogaster rDNA. *Cell.* 1977 Feb;10(2):167–76.

140. Brack C, Tonegawa S. Variable and constant parts of the immunoglobulin light chain gene of a mouse myeloma cell are 1250 nontranslated bases apart. *Proc Natl Acad Sci U S A.* 1977 Dec;74(12):5652–6.

141. Coffin JM, Welch M. 50th anniversary of the discovery of reverse transcriptase. *Mol Biol Cell.* 2021;32(2):91–97.

142. Fore J Jr., Wiechers IR, Cook-Deegan R. The effects of business practices, licensing, and intellectual property on development and dissemination of the polymerase chain reaction: Case study. *J Biomed Discov Collab.* 2006 Jul 3;1:7.

143. Hopp TP. Lymphokine racketeers? Cistron alleges foul play over IL-1, but Immunex sees it differently. *Nat Biotechnol.* 1996 Mar;14(3):275–9.

144. Marshall E. Startling revelations in UC-Genentech battle. *Science.* 1999 May 7;284(5416):885–6, 883.

145. Wiktor-Brown DM, Hendricks CA, Olipitz W, et al. Age-dependent accumulation of recombinant cells in the mouse pancreas revealed by in situ fluorescence imaging. *Proc Natl Acad Sci U S A.* 2006 Aug 8;103(32):11862–7.

146. Yengo L, Vedantam S, Marouli E, et al. A saturated map of common genetic variants associated with human height. *Nature.* 2022 Oct;610(7933):704–12.

147. Felsenfeld G, Davies DR, Rich A. Formation of a three-stranded polynucleotide molecule. *J Am Chem Soc.* 1957;79(8):2023–24.

148. Rich A. The molecular structure of polyinosinic acid. *Biochim Biophys Acta.* 1958 Sep;29(3):502–9.

149. Arnott S, Chandrasekaran R, Marttila CM. Structures for polyinosinic acid and polyguanylic acid. *Biochem J.* 1974 Aug;141(2):537–43.

150. Marini JC, Levene SD, Crothers DM, et al. Bent helical structure in kinetoplast DNA. *Proc Natl Acad Sci U S A.* 1982 Dec;79(24):7664–8.

151. Gehring K, Leroy JL, Gueron M. A tetrameric DNA structure with protonated cytosine.cytosine base pairs. *Nature.* 1993 Jun 10;363(6429):561–5.

152. Griffith JD, Comeau L, Rosenfield S, et al. Mammalian telomeres end in a large duplex loop. *Cell.* 1999 May 14;97(4):503–14.

153. Traczyk A, Liew CW, Gill DJ, et al. Structural basis of G-quadruplex DNA recognition by the yeast telomeric protein Rap1. *Nucleic Acids Res.* 2020 May 7;48(8):4562–71.

154. Chen MC, Tippana R, Demeshkina NA, et al. Structural basis of G-quadruplex unfolding by the DEAH/RHA helicase DHX36. *Nature.* 2018 Jun;558(7710):465–9.

155. Qiao Q, Wang L, Meng FL, et al. AID recognizes structured DNA for class switch recombination. *Mol Cell*. 2017 Aug 3;67(3):361–73 e4.

156. Stanton A, Harris LM, Graham G, et al. Recombination events among virulence genes in malaria parasites are associated with G-quadruplex-forming DNA motifs. *BMC Genomics*. 2016 Nov 3;17(1):859.

157. Sinclair AH, Berta P, Palmer MS, et al. A gene from the human sex-determining region encodes a protein with homology to a conserved DNA-binding motif. *Nature*. 1990 Jul 19;346(6281):240–4.

158. Ellison MJ, Fenton MJ, Ho PS, et al. Long-range interactions of multiple DNA structural transitions within a common topological domain. *Embo J*. 1987;6(5):1513–22.

159. Levens D. Cellular MYCro economics: Balancing MYC function with MYC expression. *Cold Spring Harb Perspect Med*. 2013;3(11):a014233. Epub 20131101. doi: 10.1101/cshperspect.a014233. PubMed PMID: 24186489; PubMed Central PMCID: PMC3808771.

160. Britten RJ, Davidson EH. Gene regulation for higher cells: A theory. *Science*. 1969;165(3891):349–57.

161. Horvitz HR. Worms, life, and death (Nobel lecture). *ChemBioChem*. 2003;4(8):697–711.

162. Ruvkun G, Giusto J. The Caenorhabditis elegans heterochronic gene lin-14 encodes a nuclear protein that forms a temporal developmental switch. *Nature*. 1989 Mar 23;338(6213):313–9.

163. Lee RC, Feinbaum RL, Ambros V. The C. Elegans heterochronic gene lin-4 encodes small RNAs with antisense complementarity to lin-14. *Cell*. 1993 Dec 3;75(5):843–54.

164. Höck J, Meister G. The Argonaute protein family. *Genome Biol*. 2008;9(2):210 Epub 20080226. doi: 10.1186/gb-2008-9-2-210. PubMed PMID: 18304383; PubMed Central PMCID: PMC2374724.

165. Hale CR, Zhao P, Olson S, et al. RNA-guided RNA cleavage by a CRISPR RNA-Cas protein complex. *Cell*. 2009 Nov 25;139(5):945–56.

166. Deltcheva E, Chylinski K, Sharma CM, et al. CRISPR RNA maturation by transencoded small RNA and host factor RNase III. *Nature*. 2011 Mar 31;471(7340):602–7.

167. Sapranauskas R, Gasiunas G, Fremaux C, et al. The streptococcus thermophilus CRISPR/Cas system provides immunity in Escherichia coli. *Nucleic Acids Res*. 2011 Nov;39(21):9275–82.

168. Reebye V, Saetrom P, Mintz PJ, et al. Novel RNA oligonucleotide improves liver function and inhibits liver carcinogenesis in vivo. *Hepatology*. 2014 Jan;59(1):216–27.

169. Herbert A. Flipons and small RNAs accentuate the asymmetries of pervasive transcription by the reset and sequence-specific microcoding of promoter conformation. *J Biol Chem*. 2023;299(9):105140. Epub 20230805. doi: 10.1016/j.jbc.2023.105140. PubMed PMID: 37544644; PubMed Central PMCID: PMC10474125.

170. Borchert GM, Holton NW, Williams JD, et al. Comprehensive analysis of microRNA genomic loci identifies pervasive repetitive-element origins. *Mob Genet Elements*. 2011 May;1(1):8–17.

171. Benne R, Van den Burg J, Brakenhoff JP, et al. Major transcript of the frameshifted coxII gene from trypanosome mitochondria contains four nucleotides that are not encoded in the DNA. *Cell*. 1986 Sep 12;46(6):819–26.

172. Feagin JE, Jasmer DP, Stuart K. Developmentally regulated addition of nucleotides within apocytochrome b transcripts in Trypanosoma brucei. *Cell*. 1987 May 8;49(3):337–45.

173. Shaw JM, Feagin JE, Stuart K, et al. Editing of kinetoplastid mitochondrial mRNAs by uridine addition and deletion generates conserved amino acid sequences and AUG initiation codons. *Cell*. 1988 May 6;53(3):401–11.

174. Blum B, Bakalara N, Simpson L. A model for RNA editing in kinetoplastid mitochondria: "Guide" RNA molecules transcribed from maxicircle DNA provide the edited information. *Cell*. 1990 Jan 26;60(2):189–98.

175. Lepere G, Betermier M, Meyer E, et al. Maternal noncoding transcripts antagonize the targeting of DNA elimination by scanRNAs in Paramecium tetraurelia. *Genes Dev.* 2008 Jun 1;22(11):1501–12.

176. Jahn CL, Prescott KE, Waggener MW. Organization of the micronuclear genome of oxytricha nova. *Genetics.* 1988 Sep;120(1):123–34.

177. Adleman LM. Molecular computation of solutions to combinatorial problems. *Science.* 1994 Nov 11;266(5187):1021–4.

178. Herbert A, Rich A. RNA processing in evolution. The logic of soft-wired genomes. *Ann N Y Acad Sci.* 1999 May 18;870:119–32.

179. Salvail H, Breaker RR. Riboswitches. *Curr Biol.* 2023 May 8;33(9):R343–8.

180. Stark BC, Kole R, Bowman EJ, et al. Ribonuclease P: An enzyme with an essential RNA component. *Proc Natl Acad Sci U S A.* 1978 Aug;75(8):3717–21.

181. Kruger K, Grabowski PJ, Zaug AJ, et al. Self-splicing RNA: Autoexcision and auto-cyclization of the ribosomal RNA intervening sequence of Tetrahymena. *Cell.* 1982 Nov;31(1):147–57.

182. Herbert A, Wagner S, Nickerson JA. Induction of protein translation by ADAR1 within living cell nuclei is not dependent on RNA editing. *Mol Cell.* 2002 Nov;10(5):1235–46.

183. Herbert A. Simple Repeats as Building Blocks for Genetic Computers. *Trends Genet.* 2020;36(10):739–50. Epub 20200718. doi: 10.1016/j.tig.2020.06.012. PubMed PMID: 32690316.

184. Prigogine I, Nicolis G. Biological order, structure and instabilities. *Q Rev Biophys.* 1971 Aug;4(2):147–57.

185. Turing AM. The chemical basis of morphogenesis. 1953 [Biography classical article historical article]. *Bull Math Biol.* 1990;52(1–2):153–97; discussion 119–52.

186. Gierer A, Meinhardt H. A theory of biological pattern formation. *Kybernetik.* 1972 Dec;12(1):30–9.

187. May RM. Simple mathematical models with very complicated dynamics. *Nature.* 1976;261(5560):459–67.

188. Lorenz EN. Deterministic nonperiodic flow. *Journal of the Atmospheric Sciences.* 1963;20(2):130–41.

189. Junge W. Half a century of molecular bioenergetics. *Biochem Soc Trans.* 2013 Oct;41(5):1207–18.

190. Schwille P, Frohn BP. Hidden protein functions and what they may teach us. *Trends Cell Biol.* 2022 Feb;32(2):102–9.

191. Jeffery CJ. Protein moonlighting: what is it, and why is it important? *Philos Trans R Soc Lond B Biol Sci.* 2018;373(1738).:20160523. doi: 10.1098/rstb.2016.0523. PubMed PMID: 29203708; PubMed Central PMCID: PMC5717523.

192. Gánti T. *The principles of life.* Oxford and New York: Oxford University Press; 2003.

193. Bouzon M, Perret A, Loreau O, et al. A synthetic alternative to canonical one-carbon metabolism [research support, Non-U.S. Gov't]. *ACS Synth Biol.* 2017 Aug 18;6(8):1520–33.

194. Eigen M. Selforganization of matter and the evolution of biological macromolecules. *Naturwissenschaften.* 1971 Oct;58(10):465–523.

195. Orgel LE. Self-organizing biochemical cycles. *Proc Natl Acad Sci U S A.* 2000 Nov 7;97(23):12503–7.

196. Bissette AJ, Fletcher SP. Mechanisms of autocatalysis. *Angew Chem Int Ed Engl.* 2013 Dec 2;52(49):12800–26.

197. Lederberg J, Tatum EL. Gene recombination in Escherichia coli. *Nature.* 1946 Oct 19;158(4016):558.

198. Margulis L. Genetic and evolutionary consequences of symbiosis. *Exp Parasitol.* 1976 Apr;39(2):277–349.

199. McDonald MM, Khoo WH, Ng PY, et al. Osteoclasts recycle via osteomorphs during RANKL-stimulated bone resorption. *Cell*. 2021 Mar 4;184(5):1330–47 e13.

200. Kolmogorov AN. On tables of random numbers. *Theoretical Computer Science*. 1998;207(2):387–95.

201. Turing AM. On computable numbers, with an application to the entscheidungsproblem. *Proceedings of the London Mathematical Society*. 1937;2–42(1):230–65.

202. Bao W, Jurka J. Homologues of bacterial TnpB_IS605 are widespread in diverse eukaryotic transposable elements. *Mob DNA*. 2013 Apr 1;4(1):12.

203. Jiang K, Lim J, Sgrizzi S, et al. Programmable RNA-guided DNA endonucleases are widespread in eukaryotes and their viruses. *Sci Adv*. 2023 Sep 29;9(39):eadk0171.

204. Saito M, Xu P, Faure G, et al. Fanzor is a eukaryotic programmable RNA-guided endonuclease. *Nature*. 2023 Aug;620(7974):660–8.

205. Egli M, Manoharan M. Chemistry, structure and function of approved oligonucleotide therapeutics. *Nucleic Acids Res*. 2023 Apr 11;51(6):2529–73.

206. Picardi E, Pesole G, editors. *RNA editing*. New York: Humana; 2021.

207. Fukuda M, Umeno H, Nose K, et al. Construction of a guide-RNA for site-directed RNA mutagenesis utilising intracellular A-to-I RNA editing. *Sci Rep*. 2017 Feb 2;7:41478.

208. Woolf TM, Chase JM, Stinchcomb DT. Toward the therapeutic editing of mutated RNA sequences. *Proc Natl Acad Sci U S A*. 1995 Aug 29;92(18):8298–302.

209. Monian P, Shivalila C, Lu G, et al. Endogenous ADAR-mediated RNA editing in non-human primates using stereopure chemically modified oligonucleotides. *Nat Biotechnol*. 2022 Jul;40(7):1093–102.

210. Czechowicz A, Palchaudhuri R, Scheck A, et al. Selective hematopoietic stem cell ablation using CD117-antibody-drug-conjugates enables safe and effective transplantation with immunity preservation. *Nat Commun*. 2019 Feb 6;10(1):617.

211. Herbert A, Wang AH, Jovin TM, et al. Concern over use of the term Z-DNA. *Nature*. 2021 Jun;594(7863):333.

212. Zhou Y, Xu X, Wei Y, et al. A widespread pathway for substitution of adenine by diaminopurine in phage genomes. *Science*. 2021 Apr 30;372(6541):512–16.

213. Szczerba M, Subramanian S, Trainor K, McCaughan M, Kibler KV, Jacobs BL. Small Hero with Great Powers: Vaccinia Virus E3 Protein and Evasion of the Type I IFN Response. *Biomedicines*. 2022;10(2):235. Epub 20220122. doi: 10.3390/biomedicines10020235. PubMed PMID: 35203445; PubMed Central PMCID: PMC8869630.

Index